PHYSICAL GEOGRAPHY: ITS NATURE AND METHODS

Dr Roy Haines-Young has been a lecturer in the Department of Geography at Nottingham since 1978. His research interests are in the fields of biogeography and the use of satellite remote sensing for ecological survey and monitoring. He has a long-standing research interest in the nature and methods of geography, developed in collaboration with his co-author since they were postgraduate students together. His publications include articles on the ecology of lichens, landscape ecology, and on the philosophical aspects of geography.

Dr James Petch is a lecturer in geography at the University of Salford. His specialist subjects are soil and plant ecology, hydrology particularly in relation to irrigation, geomorphology, and quantitative methods. For over ten years, with his co-author, he has developed the ideas of critical rationalist philosophy applied to physical geography. He has a particular interest in their role in the training of students. His publications include articles on pro-glacial geomorphology, irrigation management, soil reclamation and philosophical aspects of geography.

PHYSICAL GEOGRAPHY: ITS NATURE AND METHODS

R.H. Haines-Young
&
J.R. Petch

Harper & Row, Publishers
London

Cambridge
Mexico City
New York
Philadelphia

San Francisco
São Paulo
Singapore
Sydney

First published 1986

Harper & Row Ltd
28 Tavistock Street
London WC2E 7PN

British Library Cataloguing in Publication Data
Haines-Young, R.H.
   Physical geography: its nature and methods.
   1. Physical geography
   I. Title   II. Petch, J.R.
   910'.02   GB54.5

   ISBN 0-06-318327-7

Typeset by BookEns, Saffron Walden, Essex
Printed and bound by Butler & Tanner Ltd.,
Frome and London

CONTENTS

PREFACE

This book is written for students of physical geography who have done some science but little philosophy. It attempts to provide what we consider to be a much needed framework for the understanding of scientific method and its application in physical geography. It is not intended to present a review of the discipline or a history, but examples are used from a wide range of problems in order to establish that the implications apply comprehensively.

The ideas expressed here are culled from the writings of many scientists and philosophers. We have taken ideas freely just as did those from whom we took. Our hope is that others will consider what we have written, inspect it critically and take and apply what they may learn.

Although the chapters follow a sequence they are largely self-contained and therefore many important points are repeated. We see little danger in this if those for whom this is tedious can appreciate that for others the repetition of ideas in different situations adds to the understanding of the whole. Indeed this repetition will be exploited by the earnest student who can use the index and the many signposts in the text in order to explore the ramifications of major ideas. Many points have not been covered in as great a depth as we would wish, however, for reasons of economy.

We have attempted to explain the major issues through the use of case studies and exercises. The material of the case studies is discussed in the text. The exercises are more open-ended and are presented at the end of the book. They are intended to involve the student in a practical way in order that he or she can see the significance of the arguments. Readers are recommended to attempt the exercises. Most of them demand a great deal more than may appear at first sight and they should not be dismissed merely as aids to the beginner.

Many people have provided advice, criticism and other help in the writing of this book. We wish to thank John Cole and Colin Harrison who read sections of the manuscript, and students at Nottingham and Salford who attempted the exercises. Many people have contributed to our ideas. For discussion of those presented here and of our earlier works we thank Paul Mather, C.A.M. King, Rowland Moss, Ted Culling and John Gerrard. We particularly thank John Matthews who also provided material for Chapter 3 and helped in its construction. Peter Worsley kindly provided the excellent illustrations of terraces in Plate 1. The diagrams and maps were drawn by Gustav Dobrzynski, Christine Warr and Gwynneth Ashworth and we gratefully acknowledge their skill. The staff at Harper and Row have helped by their efficiency, cheerful encouragement and confidence. Finally we were blessed with intelligent, painstaking reviewers and our special thanks go to them.

CHAPTER SUMMARIES

CHAPTER 1: Logic and Judgement in Science

What is the basis of any claim that we know anything about the world? Many believe that this basis is provided by science. Scientific explanations are presented as structured arguments which arise as a logical consequence of holding some theory about the world. But why believe in any one theory rather than another? According to some people, good, rational reasons can be given for such beliefs. There are many who dispute this rationalist image of science, however. They argue that scientific knowledge is no more secure than any other form of human understanding. An analysis of these claims and counter-claims leads to an examination of the role of theories, laws and observations in scientific explanation, and a consideration of the nature of truth. The rationalist image of science is found to embody the twin ideas that the truth of a theory is judged by its correspondence to the facts, and that such judgements are based on observations of a world which exists independently of men's minds. Does this image of science as a logical ordered activity stand up to serious criticism?

CHAPTER 2: The Classical View

One very seductive view of science as a rational enterprise is the classical view. It holds that scientific knowledge is secure because it rests on experience. It involves the idea of inductive argument: that is, arguing from general statements about the world from repeated observations of particular events. Knowledge is held to be secure because of the accumulated empirical verification of ideas. This approach to science has several flaws, since there are no sound principles of verification nor of induction. Furthermore, the approach ignores the theory-dependence of observations. Observation without theory is impossible. Thus the classical approach cannot withstand criticism. The problem remains that if there is no certainty in what science does, can a rational image of science still be maintained?

CHAPTER 3: The Critical Rationalist View

Popper has shown that the flaws of the classical approach can be avoided by rejecting inductive logic in favour of deductive reasoning, and by recognizing that verification is logically not possible but that falsification is. This is the basis for critical rationalism. A rational basis for scientific knowledge is provided by deducing the consequences of theories and then attempting to expose their falsity by critical testing. If a theory survives attempts at falsification then it can be concluded only that evidence corroborates it and not that it is proved. In the process of testing, a falsified theory may be protected by changing it, but changing a theory merely to save it from refutation is bad practice. The testability or falsifiability rather than the verifiability of an idea is the criterion for whether or not it can be used in science. In practice the scientist is dealing with competing hypotheses and the problem is to distinguish between them. The real problem of testing is to find theories of greater verisimilitude. The deductive approach is altogether different from the classical tradition, however, since the rational basis for it is provided by the critical method.

CHAPTER 4: The Kuhnian View

In contrast to the classical and critical traditions of science which argue that science is essentially a rational activity, some believe that scientific attitudes and judgements are made relative to the beliefs of the scientist and of the scientific community. One such critic of the rationalist image of science is Kuhn. He views science in terms of alternating periods of stable, normal science, in which practices are dominated by some paradigm accepted by the scientific community, and periods of paradigm change. Central to it is the role of the scientific community in judging scientific work and knowledge. This is achieved within a disciplinary matrix of shared beliefs and thought exemplars or archetypal applications. These provide both the puzzles which scientists attack and the bases for judging their solutions. There is only a subjective basis for choice between competing theories or between competing paradigms. This is the thesis of relativism. This idea rests upon the incommensurability of different points of view or of different theories since if competing theories cannot be compared then the rational image of science cannot be supported.

CHAPTER 5: Lakatos and the Nature of Research Programmes

A resolution of the rationalist and relativist approaches has been attempted by Lakatos with the concept of a research programme. This is the idea that there are larger frameworks of thought than single theories, and these guide the scientist in decisions about which problems are for research and which not. A corollary is the idea that, conventionally, key assumptions in science go unquestioned. These assumptions form the 'core' of a research programme, and only secondary ideas which arise from them are open to question. This situation can explain both the use of deductive strategies with the persistence of refuted ideas, and the relativist forces which might otherwise undermine the critical approach. However, these ideas do not avoid the criticism of relativism in the operation of and in the problem of choice of

research programmes. In particular the choice between competing research programmes can only be made retrospectively. If this is so then on what rational basis
can the day-to-day problems of theory choice be made?

CHAPTER 6: Feyerabend on Method

One response to the lack of any clear method in science attributes it to the fact that
there is no scientific method. Feyerabend has argued that all attempts to identify the
methods of science are fruitless. He holds that there is no historical evidence which
demonstrates either that there is a proper scientific method or that any method has
been successful. Indeed science can be considered to advance in the face of all the
models of science currently developed. According to Feyerabend the only
methodological rule which can be defended is that 'anything goes'. Scientists, it is
argued, should adopt a pluralistic attitude incorporating any and all ideas since each
idea can potentially add to knowledge and no factual statement can provide the basis
for choice between competing theories. Above all this view of science holds that
science has no special place but is merely another of man's myths.

CHAPTER 7: The World of Ideas

Given the argument described in the previous chapters, how is one to answer the
question about how science justifies its claims to know anything about the world?
The answer can be seen to revolve around the issue of whether science can provide
good, rational reasons for choosing one theory or holding one set of observations
rather than some other. The claim of the rationalist that there is a rational basis for
scientific knowledge rests in the recognition that scientists' work results in the
growth of a body of ideas, facts, theories and opinions which exists independently of
men's minds. The content of this objective body of knowledge can be criticized and
improved by deductive reasoning and critical testing. Ideas can be compared.

In response rationalists argue that even acknowledging the role of sociological and
irrational factors, there is nevertheless no theory of incommensurability which
withstands analysis. Better tests of the rationalist view of science are provided by historical evidence. On close examination this shows the relativists' claim that the persistence of refuted ideas undermines the rationalist approach to be inadequate.
Further, 'normal' science can be seen to be no more than bad science. The fundamental
issue of these arguments is about the extent to which any of these opposing views
should be prescriptive for science rather than merely descriptive. The critical
rationalist approach, from its ability to withstand criticism and from its bases in the
actions of scientists, is held to provide the best formulated method for scientific
practice.

CHAPTER 8: Theorizing

The creation and development of theories is not susceptible to the same kind of analysis as their evaluation. In accounting for the origin of theories no source of ideas can be excluded on any rational basis. The process of theorizing can be analysed, however, in terms of the forces shaping theories and the situations in which they arise. Central to the analysis is the idea of the problem situation, which constantly evolves as theories are tested and refuted. New theories grow out of old ones since all knowledge consists in the improvement of existing knowledge. They are also influenced by the wider frameworks of thought or myths which scientists hold. Myths are some-times mistaken for theories but have a completely different role in science. Theory develops by amendment of existing theory or development of alternative ones. Amending a theory consists of changing an earlier theory in some way. In developing alternative theories any new theory is welcome since bold departures from old ideas are more likely to advance knowledge and critical science encourages the elimination of false theories. Nevertheless, there are limits to boldness. In particular new theories are rarely the product of chance thoughts but more generally of sustained commit-ment to problems.

CHAPTER 9: Modelling

Models are the devices used to make predictions. They are therefore the means by which theories can be tested. It is important to distinguish this role in science from that in engineering, where models are used to make predictions but not to test theories. Test situations are designed to allow conflict between theory and obser-vation. However, conflict may be due amongst other things to the model. Thus models as they are used in science can usefully be considered in terms of how they can cause discrepancies between theories and observations. This depends on whether they are deterministic or stochastic, whether they are hardware or software and whether they are fully or partially specified by a theory. In practice the use of models can be very complex particularly where they require calibration or optimization of parameters. In any test situation there is always the problem of judging whether or not a pre-diction is sufficiently close to an observation to reach a decision about a theory. Whatever the nature of the model and the type of prediction it provides this problem remains, as does the problem of deciding what is the source of the clash between theory and experience.

CHAPTER 10: Naming and Classifying

According to the classical view of science the naming and classification of things is an essential prerequisite to scientific study. In fact classification is impossible to achieve without some prior theory. It is undertaken in the context of some theory about how the things behave and what properties they have. To name an object and thereby specify that it belongs to a particular class, means that the properties of the class are assigned to the object. Thus names can be seen to be of little importance since they can always be substituted by a more detailed account of these properties. What is important is the set of theories which underlies the specification of the properties. The point is that since scientific classification rests on falsifiable theories there is always uncertainty in our naming of things. In practice objects are named with the degree of certainty required by the problem in hand and since theories change so do systems of classification and so do the properties or the ideas which names convey. If objects are classified on the basis of properties which are specified by non-testable theories then such objects are beyond the realm of science.

CHAPTER 11: Measuring
Measuring is the process of assigning properties to objects according to certain rules. Measurements are abstract representations of things. The amount of information which a measurement can represent depends upon the number scale being used. The significance of number scales in science relates to the precision with which predictions can be made from theories and the severity with which these theories can be tested. The way in which measurements are taken depends upon the scientific theory being tested and upon statistical theory. The scientific theory determines the measured parameter and the circumstances of measuring. The theory underlies the choice of problem and its bridge principles determine how things are measured. Statistical theory explains the relation between measurement and error. Such theories are used to determine the procedure for measuring in order that adequate measurements are made. Statistical theory also determines the repertoire of structures of relations between measured variables.

CHAPTER 12: Experimental Design
The problems of empirical science demand controlled experimentation in order to test theories. Various procedures have been termed experiments including those which set out to confirm prior expectations or merely to see what something does. Valid experiments require critical testing of stated theories either by empirical observation or by deductive analysis. The design and execution of experiments require a coordinated sequence of steps beginning with a clear statement of the problem. This statement involves the selection of the analytical method and a consideration of anticipated results. Measurements are taken in such a way as to control the effects of factors extraneous to the theory. An essential element in this control is randomization. Interpretation centres around the decision of whether or not to reject a hypothesis. Statistical theory approaches this problem through the use of probabilities but in the end it is a question of judgement.

CHAPTER 13: Physical Geography and the Critical Tradition

In the light of the arguments presented it is clear that there are problems in the methodology which physical geographers employ. The discipline lacks a critical tradition. There is a need to inject such a tradition by a conscious study of method which, it is argued, is of great practical significance. The recent history of the subject shows that what some have considered to be advances are no more than changes of fashion. They could only be considered advances if there had been developments from scientific problems through critical testing of ideas. The lack of such developments is clear. There is a need to instill the awareness of problems and to respond to them by the development of a critical tradition.

PART 1:
FRAMEWORKS

CHAPTER 1:
LOGIC AND JUDGEMENT IN SCIENCE

'How do you know?'

Science and its image
Why do the coastlines of Africa and South America look as though they could fit together like two pieces of a jigsaw? Why did almost nobody believe in the theory of continental drift before 1960, but why did almost everybody do so after 1970? How can we explain the presence of dry watercourses in arid areas? Why is Darwin's theory of evolution more widely accepted than any other? Indeed what is the basis of *any* claim that we know anything about the world? Many people believe that science provides the basis of such knowledge.

Almost everything that distinguishes the modern world from earlier centuries is attributable to science. It is easy to see the influence that science has on our everyday lives, and on our ideas about the world, but how many of us would agree that there is something special about science and the contribution that it has made? Science can easily be glamorized but is it really so different from other human activities?

People have responded in various ways to these questions about the character of science and its place in the modern world. The philosopher Karl Popper, for example, recognizes that:

> The history of science, like the history of all human ideas, is a history of irresponsible dreams, of obstinacy, and of error . . .

but concludes:

> . . . science is one of the few human activities – perhaps the only one – in which errors are systematically criticised and fairly often, in time, corrected . . . In most other fields of human endeavour there is change, but rarely progress. (1974, pp.216–217)

For Popper, it is the method of systematic criticism and correction of ideas that is the key feature which distinguishes science from all other activities. The claim that through science we have come to know something about the world is, he believes, based on the methods the scientist has used in his investigations. There are those who disagree, however. Another philosopher Paul Feyerabend, for example, has argued:

3

The reason for this special treatment of science is, of course, our little fairy-tale (the myth of method): if science has found a method that turns ideologically contaminated ideas into true and useful theories, then it is indeed not mere ideology, but an objective measure for all ideologies... But the fairy-tale is false... There is no special method that guarantees success or makes it probable. (1975, p.302)

For Feyerabend there is nothing about the methods of science which makes it more rational than other activities. The statements that science makes are no more than propaganda. Feyerabend even concludes that any attempt to identify the methods of science is fruitless.

The views of Popper and Feyerabend illustrate some of the contrasting images of science that people hold. Those who consider science to be an ordered, logical activity, with judgements based on reason, can be described as *rationalists*. Of the two contrasting opinions described above, the rationalist view of Popper probably corresponds most closely to the image of science held by the majority of scientists, but is this image correct? Can the arguments against rationalism, such as those of Feyerabend, be supported? What do these views imply about the way in which science should be practised? The purpose of this book is to look at what kind of answers can be given to these questions. Let us begin by trying to find out what scientists try to do in their work.

Case Study 1.1: What is science?
Consider the following extracts. Which of them would you regard as being 'scientific'? Try to say why. Some aspects of the Case Study will be discussed in the next section.

Extract 1   On slope terracettes in Europe
   Two major groups of theory exist concerning the genesis of these enigmatic features. One group supports the notion that terracettes are not natural slope forms but are tracks worn by grazing animals; the other supports the view that whilst grazing animals may use the terracettes, the forms themselves are due to mechanical failure in the regolith blanketing the slope ... Detailed investigation of several hundred terracettes in England, Wales and Norway by the present authors supports the view that they are due to failure. (Vincent and Clarke 1980, p.291)

Extract 2   On the relationship between river velocity and river erosion
   A river has equilibrium (stability)... when in times of greatest flood its rapidity is such that the tenacity of its bed is equal to the force and opposes the erosion not only of the bottom but also of the sides of the bed and the foot of its banks. If the velocity is too great then erosion and transport occur, and the speed is greater than that of equilibrium ... Thus by equilibrium we mean the relation between velocity of flow and resistance of bed-rock. The speed of equilibrium will vary with the nature of the bed, and the Rhône, with a bed of sizable stones, ought to have a

greater speed than that of the Meuse or Seine which flow over gravels ...
(Du Buat 1786, p.110, quoted by Chorley et al. 1964, p.89)

Extract 3 On the influence of organic debris on river channel geometry, New Zealand

Bed load transport through the study reach in Powerline Creek is highly variable in space and time and appears to be controlled to a large extent by the deposition of organic debris within the channel. A series of log and debris jams along the streams provides many temporary base levels and storage sites at which sediment can be immobilised for long periods; when such features are destroyed or disrupted bedload movement occurs as the previously stored sediments become available for transport. (Mosley 1981, pp.578–579)

Extract 4 On fog

The yellow fog that rubs its back upon the window-panes,
The yellow smoke that rubs its muzzle on the window-panes,
Licked its tongue into the corners of the evening,
Lingered upon the pools that stand in drains,
Let fall upon its back the soot that falls from chimneys,
Slipped by the terrace, made a sudden leap,
And seeing that it was a soft October night,
Curled once about the house, and fell asleep.
(from 'The Love Song of J. Alfred Prufrock', Eliot 1975, p.13)

Extract 5 On the geopolitics of Berlin

'There are many people in the world who really don't understand, or say they don't, what is the great issue between the free world and the Communist world. Let them come to Berlin. There are some who say that Communism is the wave of the future. Let them come to Berlin. And there are some who say in Europe and elsewhere we can work with the Communists. Let them come to Berlin.' (spoken by J.F. Kennedy, quoted by Miroff 1976, p.103)

Science: its aims and methods

The rationalist usually points to two characteristics of science which set it apart from other activities. These characteristics concern the aims of science and its methodology. By the term 'methodology' the rationalist simply means a way of doing things. What kinds of aims and methodology are normally identified? Let us consider first the aims of science.

The rationalist often characterizes the aim of science as 'the discovery of true theories' or 'making accurate predictions or observations about the world'. If you agree that science has such aims then you might have considered that the first three extracts of Case Study 1.1 are 'scientific'. To the extent that politicians also attempt to make true statements then, on these grounds, you would also have to include the last extract about Berlin. The poet, on the other hand, does not seem to be constrained by 'truth' in quite the same way. Eliot creates an interesting image of the animal-like behaviour of fog, but the acceptability of his ideas is not judged according to whether fog does or does not behave in the way described. Ideas about truth and falsity simply do not enter into the matter.

The rationalist would argue that attempting to discover true theories or make accurate predictions and observations is part of a more fundamental aim, that is to *explain* how the world works. Although science is not unique in having this aim, what is supposed to give science its special place is the way it tries to achieve this end, that is its methodology.

Science, like other human activities, involves making judgements about ideas. For some scientists, judgements may concern the choice between conflicting theories, as in Extract 1 of Cast Study 1.1. For others they involve testing the accuracy of predictions, such as those made by Du Buat in Extract 2, or the reliability of observations, such as those of Mosley in Extract 3. Whatever the situation, the rationalist assumes not only that there is some rational, logical basis which guides these judgements but also that there is an external reality against which the ideas themselves can be tested. Theories, predictions and observations are accepted or rejected according to the way in which they appear to match a reality which is, in principle, accessible to each of us.

Despite the persuasiveness of President Kennedy's words, and despite his attempt to speak of the world as he knew it, his ideas cannot be assessed in the same way as we would assess those of Extracts 1, 2 and 3. By describing the first three extracts as 'scientific' what is meant is that judgements can be made about them which are based either on independent observations or on predictions which would confirm or deny them. We would also hope that even the authors of the passages in question would have this attitude. Anybody can form a judgement about these extracts and the presumption is that eventually such judgements would coincide because there is a single external reality which we seek to understand. Although Kennedy's words describe a reality, the judgement we make about them does not ultimately depend on anything observable and external to us. There is nothing we can point to which would allow us to judge whether such ideas were 'right' or 'wrong'. The judgements we form about his ideas depend more on the set of values or opinions which we hold. Kennedy gave the speech from which Extract 5 was taken on his visit to Berlin. It was prompted by the contrast he perceived between the two political systems which stood in opposition to each other on either side of the Berlin wall. The reality he saw, however, cannot be used as a basis for judging his ideas. Imagine that the words had been spoken by Kruschev, the contemporary Soviet leader, and ask yourself what meaning they then would have conveyed. The meaning of the passage depends on who is speaking and not on the thing of which he speaks. Our judgement about the statements is based on the values we attach to the different political systems. The choice between such systems is not the same as the choice between competing theories. Within the bounds of what is physically possible, political systems can be organized in a number of ways. The choice between them is not constrained by any external reality.

A similar kind of argument might be put forward for excluding Eliot's account of fog from science. Judgements about the passage cannot be made by taking further observations of the world. Our judgement, in this case, is based more on aesthetic and linguistic values.

Thus the methodology which the rationalist identifies as making science distinctive is one which allows the scientist to use experience as the basis for making rational judgements about ideas and explanations. The aim of science is to solve problems and to explain things about the world, and ideas are judged according to whether they match a reality that is independent of the human mind.

Does this rationalist image of science stand up to serious criticism? In order to answer this question we must first consider in more detail what the act of scientific explanation involves.

Explanation in science

Explanation has been described as 'the process of making things intelligent', or of reducing the unfamiliar to the familiar (Lacey 1976; Hospers 1980a). More generally, explanation can be thought of as simply a statement of why things are as they are. According to the rationalist, what distinguishes scientific explanation is its relationship, on the one hand, to the world of ideas, and, on the other, to the world of observable things.

In scientific work explanation usually involves saying why objects or events, whose character we find puzzling, appear or behave as they do. Their unfamiliarity is dispelled by showing that their character or behaviour is part of the general structure and function of the world. The puzzling objects and events are simply manifestations of these underlying patterns and processes. How do we go about constructing such explanations?

Explanations arise as a necessary consequence of holding some view or theory about the way in which the world works. This means that the theory provides reasons for things and an explanation is what follows logically from it. We can say 'an event occurs or an object behaves as it does because . . .' and the reasons provided by the theory then follow. Thus explanations are linked to the world of ideas (theories and hypotheses) by logical (deductive) reasoning. In contrast, explanations are linked to the world of observable things by the conceptual frameworks provided by theories. Things are explained in a certain way because the scientist holds particular ideas about how the world is structured or how it behaves. But why choose one theory about the world rather than another? If 'good reasons' can be found for choosing one theory rather than another then one explanation can be preferred to another. According to the rationalist there are ways in whch the scientist can gain access to the world of observable things in order to make such judgements about the acceptability of competing theories. Explanations are, therefore, related to the world of observable things by theories and the judgements which scientists make about these theories.

These ideas about scientific explanation may seem complicated and so before they are discussed in detail let us illustrate the character of scientific explanation by means of a case study. The study concerns the problem of arroyos.

Case Study 1.2: The origin of arroyos

Arroyos are gully-like landforms which occur in arid and semi-arid areas of the world. Cooke and Reeves (1976) describe such features as being common throughout the American South-West. They are cut into the debris deposits that cover valley floors or which form otherwise undissected plains (Figure 1.1). Arroyos are generally characterized in terms of their steeply sloping or vertical walls in cohesive, fine sediments, and by their flat and generally sandy floors. Arroyos carry water flow only intermittently. Their origin has been the focus of interest for a number of years. Rich, for example, wrote in 1911:

> A conspicuous development of recent stream trenches in the valleys of many of the temporary streams of the western states is a feature of such widespread and common occurance that it cannot be assigned to accidental causes, but *calls for explanation which shall have more than a local application* ... (p.237, our italics)

What sort of explanation might an earth scientist devise for the features described by Rich?

Graf (1979) suggested that arroyo formation *begins* when the tractive force exerted by flowing water exceeds the threshold resistance of material of the valley bottoms. The threshold resistance depends on the character of the substrate materials and their vegetation cover. In turn, the tractive force of flowing water depends upon the magnitude and frequency of precipitation events to which an area is subjected. A change in either threshold resistance or tractive force can cause arroyo formation to begin.

The ideas of Graf (1979) are of a very general kind and on their own it is difficult to see how they could explain every arroyo. How could these ideas be used to explain the particular arroyos for the area that Rich was interested in, namely part of the Silver City Quandrangle of New Mexico?

Ideas such as those of Graf can be made to apply to the area in question if they are considered in relation to statements about the circumstances that have applied in the Silver City Quandrangle. Suppose that investigations in the area have shown that vegetation cover had been reduced at some time in the past as a result of overgrazing. Indeed, this is what Rich (1911, p.242) reported following local interviews. Using such data, an explanation of arroyos in the Silver City Quadrangle of New Mexico might run as follows:

1. Arroyo cutting begins whenever the tractive force of flows exceeds the

Figure 1.1 An arroyo in the Silver City Quadrangle, New Mexico (based on a
   photograph of a stream trench of the Mangas Valley by Rich 1911)

threshold resistance of materials (where resistance is a function of vegetation cover).

2. Vegetation cover in the Silver City Quadrangle of New Mexico has been reduced by overgrazing since the 1870s such that the threshold resistance of substrate materials was lowered below the critical level for erosion.

3. Arroyos have developed in the Silver City Quadrangle of New Mexico since the 1870s as a result of overgrazing.

The final statement is divided from the other two by a line to indicate that it represents a conclusion of some kind.

To what extent does Statement 3 follow on directly from the other two?
Do Statements 1 and 2 completely determine 3 or only provide a partial justification of it?
Is the explanation of arroyo formation given above acceptable? Has it made arroyos 'intelligent'?

The explanation will be discussed in more detail below. At this stage make a note of any points that might stand in its favour as a good explanation or any points which might undermine it. Try Exercise 1.1

The term *deductive* is often used to describe the form of the explanation illustrated by the example of arroyos because of its *logical structure*. The explanation is logical in the sense that it is made up of a set of consistent and related statements which are compatible. The origin of arroyos in the Silver City Quadrangle is explained by showing that their formation follows logically (that is, it can be deduced) from two sets of statements, namely the kind of general statement about arroyos given in 1, and the more specific observations contained in 2.

The various parts of a deductive argument such as the one shown above can be described by different technical terms. Statements 1 and 2, for example, are sometimes known as the *premises* of the explanation and together they form the *explanans*. The thing explained (e.g. the origin of arroyos) is described as the *explanandum*. Such terminology need not detain us here, however. The most important thing to note is its general form, which can be regarded as an example of the *covering law model* of explanation (sometimes also known as the deductive-nomological model). In the covering law model of explanation Statement 1 is a *covering law* and Statement 2, about the particular circumstance of the event being explained, is known as the *initial conditions*. The covering law and the statement about initial conditions logically entail the conclusion which represents the explanation. If Statements 1 and 2 are true then the conclusion, 3, is true. If an explanation has this form then it is said to be *deductively valid*.

The premises of a deductive argument are statements which in the scientist's judgement describe the way in which the world is actually structured. The conclusion of the argument, which represents the explanation, is a logical consequence of holding such beliefs about the world. Thus the connection between the explanation and the world of ideas is a logical one. The link between the explanation and the world of observable things is provided by the judgement that the premises of the argument correspond to the way things actually are. Thus if we regard the explanation about arroyos as acceptable it is on the basis of the judgement about the correspondence of the underlying theory to reality.

In order to become more familiar with the covering law model and the character of valid explanations consider Case Study 1.3.

Case Study 1.3: The covering law model
Consider each of the following questions. Which of the explanations offered are based on deductively valid arguments? Once again we will consider the material in the next section.

(i) Why are the basalts of Europe more susceptible to frost shattering in cold maritime climates than in cold continental ones?
   1. The rate of frost shattering is directly proportional to the number of freeze–thaw cycles per unit time.
   2. The basalts in the cold maritime areas of Europe are subject to more freeze–thaw cycles per unit time than are those of the cold continental areas.

3. Basalt weathers more rapidly by frost shattering in cold maritime climates than in cold continental ones, because of the greater number of freeze–thaw cycles per unit time in maritime areas.

(ii) Why did the fish die in Lake Erie?
   1. The probability of death following severe mercury pollution is high.
   2. Lake Erie suffered severe mercury pollution.

3. The fish in Lake Erie died as a result of severe mercury pollution.

(iii) Why did it rain?
   1. All rain is caused by witch doctors dancing.
   2. A witch doctor danced here today.

3. The witch doctor made it rain by dancing.

(iv) Why did the River Thames form terraces?
   1. Glaciation occurred because of reduced net global radiation.
   2. Sea level was lowered in a step-wise fashion as a result of glaciation.

3. The River Thames formed terraces because of a reduced base level for erosion.

(v) Why do birds fly?
   1. The purpose of all evolutionary adaptions is to ensure survival.
   2. Birds which fly show higher survival.

3. Birds evolved flight in order to survive.

Our account of explanation in science has been simplified because our aim has been to introduce the general problem. It might seem, for example, that we have been extremely naive in suggesting that all explanation in science is deductive in form. Those familiar with the scientific literature might point to apparently quite distinct types of explanation. Nagel (1961), for example, has described at least three other possibilities:

1. *Probabilistic explanations:* these are similar in form to the covering law model except that the law-like statement contained in the explanation only refers to some statistical or probabilistic relationship. Example (ii) in Case Study 1.3 illustrates an explanation of this type.
2. *Genetic or historical explanations:* these seek to explain an event or state of affairs by reference to the sequence of events which have led to it, as in Example (iv) above.
3. *Functional or teleological explanations:* these seek to explain something by reference to their purpose. Such explanations are commonly found in the biological sciences and are illustrated by Example (v) above.

While these, and no doubt other types of explanation might be identified, it is a matter of some argument as to whether any but the deductive form constitutes an acceptable explanation at all. This is because with all but the deductively valid form, the conclusion does not follow of necessity from the premises of the argument. If the premises of the deductively valid explanation are true then the explanation is true. In order to illustrate the drawbacks of the other kinds of explanation let us consider the examples of Case Study 1.3 in more detail.

Setting aside the problem of whether any of the explanations offered in Case Study 1.3 are true or not, only Examples (i) and (iii) have the property of being deductively valid. Their conclusion is inescapable given the premises of the argument. By contrast, in the case of the probabilistic explanations, the conclusion does not necessarily follow from the premises. In Example (ii) Statements 1 and 2 can be true but 3 can be false and there would be no contradiction. This is so because the first premise is not of an 'all or nothing' kind like the first statements in Examples (i) and (iii). If the conclusion does not follow logically from the premises then it is difficult to see what claim the conclusion has for being a coherent explanation. The explanation is not an inevitable consequence of holding some theory about how the world behaves. Similar problems arise with the historical explanation. In Example (iv) the explanation consists of a description of a sequence of events and it does not necessarily follow that terraces will always be formed in the way suggested. A different sequence of events would provide an equally valid explanation and there is no way of deciding between them. This conclusion would apply regardless of how long the list of historical observations was made. The teleological or functional explanation illustrated by Example (v) is also suspect, although the reaons for doubting it are more subtle. The argument appears to be deductively valid but it assumes goal-directed behaviour. So far as we are aware, only man is capable of conscious, goal-directed behaviour. Moreover, the discovery of 'purpose' is, in any case, fraught with difficulties. Purposes can only be recognized in terms of outcomes and there is no way in which we can gain independent evidence of their existence. Purpose does not correspond to anything real that can be measured and so the explanation is an empty one.

The reasons for rejecting teleological arguments as providing an acceptable way of explaining things are important because they begin to focus our attention on the essential properties of a valid scientific explanation. Although we have argued for the necessity of basing explanation on deductively valid arguments, this must not be taken to imply that all deductively valid explanations are satisfactory scientific explanations as may be illustrated by reference to Examples (i) and (iii) in Case Study 1.3. Both sets of statements are deductively valid but clearly Example (i) seems more respectable than (iii). However, Explanation (i) is no more plausible than (iii) if we confine ourselves to logical considerations. What is it about (i) that might make it a valid explanation over and above matters of logic? The answer is to do with the truth of the premises which make up the argument.

The acceptability of a deductively valid argument as a proper explanation cannot be decided only by attention to matters of logic. The property of being a deductively valid argument is a necessary condition of a valid explanation but it is not a sufficient one. Once logical matters have been settled the acceptability of the explanation depends on the extent to which the scientist has been successful at identifying those laws and initial conditions which correspond to the way in which the world actually works. It is at this point that scientific judgement enters. Explanations are linked to the world of ideas by logic, and arise because we hold certain views about how the world is organized. Such views are embodied in the theories, laws and observational reports which in the scientist's judgement describe reality. It is by such judgements that the explanation is linked to the world of observable things. For the rationalist, the discussion of scientific method involves giving an account of the way in which these judgements about theories, laws and observational reports are made. Given such a view one may ask whether the premises of a deductively valid argument can ever be held to be true, or whether rational reasons can be given for maintaining one set of premises rather than another. These questions are fundamentally important and the kind of answers that can be given will form an important element of the remaining chapters of the first part of this book. In order to provide the basis of this dicussion we must first look at the character of scientific laws and theories in more detail.

Laws in science

Statements which scientists call 'laws' are so general in character that they are often described as *universal statements*. They describe some characteristic or behaviour of all the members of a class of things. Laws are not, however, used to provide definitions. Instead they tell us something about a thing that would not usually be employed in recognizing it. Universal laws assert that some relationship holds at any time or any place as a consequence of the way in which the world is structured. Laws describe processes, relationships between things, or behaviour that is part of the very fabric of the universe (Hospers 1980b).

These ideas will probably seem very abstract. In order to explore them in more detail consider the following case study.

Case Study 1.4: Scientific laws
Which of the following statements would you consider to be a scientific law? Try to say why.

1. All limestone contains $CaCO_3$.
2. Complete competitors cannot coexist.
3. Stream junctions are accordant.
4. All dragons breathe fire.
5. Freely moving bodies are deflected to their right in the northern hemisphere and to their left in the southern hemisphere.
6. For oceanic islands: $\log_{10} S = 0.18 \log_{10} A + 1.26$, where S is the number of bird species per km^2 and A is area in km^2.
7. All the tors on Dartmoor are composed of granite.
8. No vehicle in Los Angeles emits more than 3.4 gm CO per vehicle mile.
9. All mammals are terrestrial.

In evaluating the statements in Case Study 1.4, it is worth noting at the outset that there is no simple formula that can be applied to determine whether or not a statement qualifies as a scientific law. The general form of a scientific law can be described as follows:

For any x, if x is an A, then x is a B.

However, merely because a statement conforms to this pattern it does not follow that it is a law in the generally accepted sense.

Consider Statement 1 about limestone. It appears to correspond to the formula: for any x (where x is a rock) if x is a limestone then x is calcium carbonate. You may be reluctant to call Statement 1 a scientific law, however, on the grounds that it really tells us no more than we already know once we are aware that the rock we are dealing with is a limestone. Rocks cannot be limestone and not contain calcium carbonate. This illustrates the point that laws must extend our knowledge about things and not merely define them. Thus in the formula given above, A must not simply follow on logically from B. A test of this property is that if you deny B does it automatically follow that A is false? If it does, then the statement is not a scientific law.

The reason for rejecting the statement about limestone as a law tells us something important about the general character of scientific laws. That is that the law tells us that certain properties are associated or connected with each other and that this relationship is not merely a logical connection. What is it about this relationship which makes the statement a law? Is it to do with the truth or falsity of the statement?

Unfortunately, while we require scientific laws to be true it does not follow that all true statements are laws. To illustrate this point consider Statements 9, 4 and 7. Statement 9 about mammals might be rejected as a scientific law simply because it is false. Whales are mammals and they are aquatic. However, Statement 4 is true but it is hardly a law of nature. Since dragons do not exist it is simply vacuously true. It is as true to say all dragons breathe fire as to say that they do not. Similarly, if the accounts of the many geomorphologists who have visited Dartmoor in England are to be believed, then Statement 7 is true but, once again, it is not a law of nature. In this case you might reject it on grounds that it asserts no more than an *accidental* relationship between two things. Although all tors on Dartmoor are composed of granite, would you be prepared to assert that all tors, anywhere at any time, are composed of granite? You may be unwilling to say 'yes' because you know that tor-like landforms can occur in limestone and sandstone areas. Statement 7 illustrates the important point that scientific laws describe a relationship between things which is not merely a logical or accidental connection, but one which is a true and *necessary* association.

The idea of necessity can be explained more fully by reference to Statement 5. It is clear that Statement 5 conforms to the formula for a scientific law: for any x, if x is a freely moving body, then x will be deflected to the right in the northern hemisphere and to the left in the southern hemisphere. Moreover, the statement is universal, and the property of being deflected in the way described is not one that we would use to define a freely moving body. On the basis of our experience, at least, we might also be prepared to accept that the statement is true. In addition to these properties, however, the feature which might make us inclined to consider it as a scientific law is that we would regard it as physically impossible for freely moving bodies to behave in any other way than in the one suggested. The relationship described is not merely accidental but physically necessary. This additional characteristic of a scientific law is sometimes described as *nomic necessity*, and is the key feature about which judgement has to be made when deciding the status of a scientific law. Nomic necessity is important because only if a statement has such a property can it be said to be a law and provide an adequate basis for an explanation.

The concept of nomic necessity can be illustrated further by a comparison of the reasons for accepting Statement 5 as a law but rejecting Statement 8. All of the cars in Los Angeles may well emit no more than 3.4 gm carbon monoxide per vehicle mile. They may do so because they are physically incapable of generating more than 3.4 gm carbon monoxide per vehicle mile by virtue of the way in which they are constructed in accordance with antipollution laws. We would not argue, however, that the relationship contained in Statement 8 is one of nomic necessity. The statement does not describe anything that we would regard as an inherent property of nature. The emission levels set are arbitrary, and emission levels could be changed by modifying the construction of cars.

The idea of nomic necessity might at first sight seem a rather elusive concept. This is so because there is no simple rule which can be used to tell us whether or not a law-like statement possesses this property. The assertion that a statement has the property is based on the judgement of the scientific community formed in the light of observation and experiment. A problem faced by physical geographers is whether any of the relationships noted in their study of the world has the property of nomic necessity (cf. Guelke 1971). The issues may be illustrated by a discussion of the remaining statements in Case Study 1.4.

Statement 2, for example, about competing organisms is sometimes known as Gause's principle. It suggests that no two species can occupy the same niche and coexist in the same place at the same time. One will always exclude the other as a result of competition. Although ecologists have not gone so far as to call it a law, some regard it as one of the basic assumptions of community ecology (Hardin 1960). Other workers (e.g. Cole 1960), however, have suggested that it is a 'trite maxim' and have rejected any claim that it is a fundamental law or principle. Such arguments need not concern us here. The point which such a discussion illustrates, however, is that a statement is not regarded as a scientific law on the basis of logic alone. Judgement has a key role to play.

A similar kind of judgement must be made about Statement 6. Although this statement differs from the others in that it expresses a quantitative relationship, it nevertheless conforms to the pattern of scientific laws. It asserts that for any island, if the island has an area A, then the island has S bird species. Whether the equation is regarded as an expression of a scientific law, rather than some accidental empirical association, depends upon whether the relationship is part of some more all-embracing body of theory that sets out the nature of the causal connection. Whether we can agree that the species–area relationship is a 'fundamental law of community ecology' (Schoener 1976) depends on the view one takes of the theory of island biogeography (cf. Gilbert 1980; Diamond and May 1981).

Statement 3 in Case Study 1.4, is a summary of Playfair's law which can be expressed more fully as follows:

> Every river appears to consist of a main trunk, fed from a variety of branches, each running in a valley proportional to its size, and all of them forming a system of valleys communicating with one another, and having such a nice adjustment of their declivities, that none of them join the principal valley, either on too high or too low a level; a circumstance which would be infinitely improbable if each of these valleys were not the work of the stream which flows in it. (Playfair 1802, p.102, cited by Chorley et al. 1964, p.62)

The physical geographer Barbara Kennedy (1984) has reported observations on the accordance of stream junctions. It is the judgement that we form of such work that determines the view we take of the validity of Playfair's law. Kennedy suggests that while geomorphologists since Playfair have apparently accepted that fluvial confluences will be accordant, in distinction to the hanging junctions of glacial troughs, her field observations have identified a

> ... wide range of circumstances in which discordant (i.e. hanging) confluences could be produced in fluvial networks. (p.155)

These observations may appear to contradict Playfair's law but the judgement about its validity is not so simple. There are two problems to be faced. First, it is not clear whether Playfair was dealing with rivers or valleys. Kennedy seems to place more emphasis on rivers, whereas the case can be made that Playfair's law only relates to the accordance of valley junctions. The relevance of Kennedy's work can only be judged on the basis of our interpretation of what Playfair intended. Second, there is the problem of the scale at which Playfair's law is supposed to operate. Was Playfair's law meant to apply at the physical and temporal scales at which Kennedy's work was conducted? Playfair seemed preoccupied with establishing the validity of the idea of fluvially eroded landscapes formed over long periods of time in contradistinction to the prevailing ideas of the catastrophists. He was not, like Kennedy, concerned with the dynamics of open channel flow. At the level at which Playfair wrote it might be that his law is not refuted by Kennedy's observations. This is a matter for judgement.

We argued in the first section of this chapter that the study of scientific method is important because it is about the way in which judgements are formed in science. These judgements often deal with assessing the validity of competing explanations about the way the world works. From the discussion presented above it is clear that often these judgements concern an assessment of the status of scientific laws which are used in explanations. It is clear from the examples, however, that these judge-ments about laws, and therefore about explanations as a whole, cannot be taken in isolation from other scientific work. The examples discussed illustrate that decisions can only be made by looking at the scientific context in which the laws are set. We have described this context quite casually in terms of 'scientific theories' and 'empirical observations'. In order to gain a deeper insight we must now investigate the character of theories and empirical observations a little further.

Theories and hypotheses

When a scientist has to decide whether an explanation is acceptable or not, rarely can the decision be made on the basis of logical reasoning alone. As we have seen, deduct-ively valid arguments must be part of scientific explanation but they are not the only part. Scientific explanations must also be judged on the basis of the *truth* of their premises. These judgements are made from the background of knowledge and experience which the scientist possesses. These background ideas are described as scientific *theories* or scientific *hypotheses*. In the account which follows we will use the terms hypothesis and theory interchangeably although some reserve the word theory for a set of ideas which are 'better established'.

The ideas contained in a theory or a hypothesis represent a framework which can be used to explain the regularities that are observed. These regularities are expressed in terms of scientific laws. The theory explains these laws by describing the mechanisms or processes which lie behind them. The theory shows why the relationship expressed by the law is one of nomic necessity. Although theories are general in character, they are more than abstract accounts of the world. Good scientific theories must describe how abstract entities behave and show how they correspond to things that can be observed and measured. Hemple (1966) has described these two elements as the *internal principles* and the *bridge principles* of a theory. Their character will be illustrated by continuing our consideration of the origin of arroyos (Case Study 1.2).

Case Study 1.5: Theory and measurement
Consider Hemple's distinction between the internal principles and the bridge principles of a theory which is described above, and then study the extract which follows. Try to identify those sections of the account which describe the general processes involved in arroyo formation, and those sections which tell us how the effects of these processes can be observed and measured.

Prior to arroyo formation, the relevant sections of valley floors were evidently either relatively stable or aggrading. Many apparently possessed no channels at all. These are two fundamental changes that – together or individually – could have led to the localized erosion and arroyo formation: increased erodibility of valley-floor materials, and increased erosiveness of flows through valley bottoms.

Erodibility, a measure of the ability of sediment and soil to withstand erosion forces, is largely determined by particulate and structural properties of the materials (cohesiveness, permeability and partical shape, size and density, etc.) and by the extent to which the material is protected from the direct attack of erosion processes. In the valley floors of the South-West, increased erodibility is likely to result from one or more of three changes. Vegetation changes – reduction, removal or replacement – might increase the extent to which materials are directly exposed to the action of running water. Weakening of soil structure could result directly from cultivation, from trampling of animals and passage of vehicles, or indirectly from vegetation changes. Those activities causing compaction of soil could also lead to the reduction of infiltration capacity in the soil, thus increasing in certain circumstances the availability of surface water. Finally, the masking of valley-floor vegetation and soil by flood sediments . . . could provide more erodible surface conditions, thus facilitating entrenchment . . .

Erosiveness, the propensity for flows to detach and remove materials from a given reach more rapidly than they are replaced from upstream, is a function of numerous interrelated hydraulic, channel-form, and sediment-load variables, and is exceedingly difficult to evaluate precisely. However, several relevant approaches to measuring erosiveness have been devised over the years by engineers in their attempts to design stable channels in alluvial materials . . . One of the oldest and most easily comprehended of these, *the method of permissible velocity*, uses mean velocity of flow as a surrogate for erosiveness. (Cooke and Reeves 1976, pp.15–17)

One of the major points made in Case Study 1.5 is that arroyo formation depends on the balance struck between erodibility of valley-floor materials and the erosiveness of potential flows. 'Erodibility' and 'erosiveness' are general concepts, and the statements about them represent the more important *internal principles* of the theory.

Erodibility is a measure of the ability of materials to withstand erosion, while erosiveness is the propensity of flows to detach and remove materials. Although they are properties of sediments and fluvial systems respectively, neither is directly measurable. They are, however, the kinds of property that might appear in a general law. Consider, for example, the law-like statement in the explanation of arroyos given in Case Study 1.2:

> Arroyo cutting begins whenever the tractive force of flows exceeds the threshold resistance of materials.

It is easy to see that 'tractive force' of flows and 'threshold resistance of materials' correspond to the ideas of erosiveness and erodibility. The 'law' simply states the kind of relationship that exists between them and how this is responsible for arroyo formation. On their own, however, such statements cannot explain the occurrence of any particular arroyos since they do not specify how erosiveness and erodibility can be observed in the field. This additional element is provided by the *bridge principles* of the theory.

Some of the more obvious bridge principles of the theory outlined above are the statements describing how erosiveness is to be measured. Cooke and Reeves note that it is a property which it is difficult to measure directly and suggest that it may be done indirectly using 'the method of permissible velocity'. The theory asserts that mean velocity of flow can be regarded as a *surrogate* for erosiveness. To say that mean velocity is a surrogate for erosiveness is the same as saying that the actual measurement of mean velocity is judged to stand in for the measurement of erosiveness. The bridge principle is clearly important because it specifies how an abstract property can be investigated by assessing other properties of the fluvial system which are tangible and observable. In a sense the abstract ideas represented by the internal statements of a theory are attached to reality by the associated bridge principles.

Although the extract mentions that erodibility depends on such factors as particle size distribution and permeability, the bridge principles which tell us how erodibility can be assessed are different in character from those about erosiveness of flows. The bridge principles relating to erodibility focus on what kinds of process can cause it to change. One such process that can cause erodibility to change is vegetation change. Since plants provide protection to a surface, a change in vegetation cover must affect erodibility. Thus, once again a judgement is made that the behaviour of an abstract property of the system is measured by using some surrogate.

Just as the internal principles of the theory formed the basis of the law-like statement in the explanation of the origin of arroyos, so the bridge principles form the basis of statements about initial conditions. The statement that 'arroyo cutting begins whenever the tractive force of flows exceeds the threshold resistance of valley floors' can only explain the origin of arroyos at a particular place because the bridge

principles tell us that changes in erodibility are to be measured by changes in vegetation cover. Bridge principles, therefore, not only tie a theory to reality, but also allow
explanations of particular phenomena to be developed in terms of general
theories.

Despite the risk of some repetition, Case Studies 1.2 and 1.5 have focused on the
same problem. Our aim in selecting a single theme is to illustrate the kind of
relationship that exists between explanation and the background body of theory on
which it is based, and thus that science is not a mechanical pursuit, and judgements
about the acceptability of competing arguments and explanations cannot be decided
by the application of any simple set of rules. Jugements are shaped according to the
view which the scientist takes of scientific laws and the theories or hypotheses on
which they are based, together with the observational and empirical evidence relevant to the problem. The scientist chooses to base explanation on one theory rather
than on another because of the judgement he makes about competing ideas. The way
in which experience can be used to provide the basis of such judgements is a major
problem for the rationalists. When explanations are challenged the rationalist claims
that there are good, rational reasons for explaining the world in terms of one theory
rather than another. Some critics of rationalism argue that no such basis exists. The
question of whether the image of science put forward by the rationalists can be
supported will form the major concern of the remaining chapters in Part 1 of
this book.

Science, truth and realism

The claim that scientific judgements and beliefs have a rational basis has implications for ideas about 'truth' and 'reality'. We must say a little about these before we
look at the arguments for and against rationalism.

Science is not unique among human activities in its aim to describe and explain the
world. As the extracts in Case Study 1.1 show, the poet and the politician might also
have such aims. A feature that distinguishes science, however, is the way it tries to
provide a *true* account. Theories are not accepted or rejected on grounds of their
aesthetic appeal or that they are inspiring. Observations about the world are not
reported merely because the scientist likes the view. Rather, science aims to explain
and report the world *as it is*, and it is on this basis that scientific work is judged.

The idea that science tries to understand the world as it is may seem fairly obvious,
as is the implication that the scientist must suppose that there is a tangible world to
be understood. However, such a *realist* view is by no means universally accepted by
all those who have tried to give an account of scientific methods. The character of the
opposing schools of thought may be seen by looking at the different ideas which are
held about the nature of 'truth'.

Newton-Smith (1981) suggests that realism has been used to characterize many
different positions in the philosophy of science. All, he notes, involve the assumption
that scientific propositions are regarded as true or false by virtue of how the world is

structured *independently of ourselves*. The assertion that statements are either true or false according to whether or not they coincide with patterns in the real world is known as the *correspondence theory of truth*. This theory deals with the relationship between that which is true (namely a statement, a belief or a judgement) and that which *makes* it true (a fact, a state of affairs or an event). A statement is true or false if, and only if, it corresponds to the facts. The geologist Hutton seems to have implied something close to the correspondence theory when he wrote:

> In examining things which actually exist, and which have proceeded in certain order, it is natural to look for that which has been first; man desires to know what has been the beginning of those things which now appear . . . A theory of the Earth, which has for object truth, can have no retrospect to that which has preceded the present order of the world . . . A theory, therefore which is limited to the actual constitution of this earth, cannot be allowed to proceed one step beyond the present order of things. (1795, I, pp.208–281, see Chorley et al. 1964, pp.36–37)

> It is the philosophy of nature that the natural history of the earth is to be studied; and we must not allow ourselves ever to reason without proper data, or to fabricate a system of apparent wisdom in the folly of a hypothetical delusion. (1795, II, p.564, see Chorley et al. 1964, p.37)

In the extracts Hutton clearly implies that there is an external world against which ideas are assessed. Since the correspondence theory asserts that there is an independent reality against which the truth may be judged, this theory entails realism.

The correspondence theory and realism seem straightforward and obvious. Certainly they are beliefs implicit in much scientific work. They appear in various guises in what we will call the 'classical tradition' (Chapter 2) and in 'critical rationalism' (Chapter 3). There are, however, alternative views. In opposition to the correspondence theory, there are two other ideas about truth which it is important to mention, namely the *coherence theory* and the *theory of pragmatic utility*.

The coherence theory of truth asserts that a statement is true if it is logically consistent (or coherent) with the rest of one's beliefs. According to the coherence view what is true can change as one passes from society to society or from one theory to another. Consider the following extract from the writings of the seventeenth-century astronomer Francesco Sizi arguing against Galileo's claim that as a result of his telescope observations moons could be seen circling Jupiter:

> There are seven windows in the head, two nostrils, two ears, two eyes and a mouth; so in the heavens there are two favourable stars, two unpropitious, two luminaries, and Mercury alone undecided and indifferent. From which and many other similar phenomena of nature such as the seven metals, etc., which it were tedious to enumerate, we gather that the number of planets is necessarily seven . . . Moreover, the satellites are invisible to the naked eye and therefore can have no

influence on the earth and therefore would be useless and therefore do not exist. (see Holton and Roller 1958, p.160)

For those, like Sizi, who maintain the coherence theory, there is no notion that truth is judged against an external reality in the way envisaged by the correspondence theory. Instead, truth is regarded as something which is relative to the social perspective or current beliefs of the scientist. People who hold such a view of truth can be described as *relativists*. A further illustration of the relativist view is provided by the Reverend Gerald Melloy in his conclusion to *Geology and Revelation: or the Ancient History of the Earth Considered in the Light of Geological Facts and Revealed Religion*:

> We have, then, two distinct systems of interpretation, according to which the vast Antiquity of the Earth, asserted by Geology, may be fairly brought into harmony with the history of Creation, recorded in Scripture. The one allows an interval of incalculable duration between the creation of the Heavens and the Earth, and the work of the Six Days: the other supposes each of these six days to have been itself an indefinite period of time . . . neither is at variance with the language of the Sacred Text . . . Either mode of interpretation seems itself quite sufficient to meet all the present requirements of Geology; for, according to the Bible narrative, either interpretation would allow time without limit for the past history of our Globe; and time without limit is just what Geology demands. (1873, p.428)

Although we may now view the arguments of Sizi and Melloy as absurd, the thesis of relativism which they illustrate is one which continues to be argued forcefully. A modern-day expression of the view appears to be provided, for example, by Haggett and Chorley in their account of models:

> Models and theories are very closely linked . . . perhaps only differing in the degree of probability with which they predict reality. The terms 'true' and 'false' cannot usefully be applied in the evaluation of models, however, and must be replaced by ones like 'appropriate', 'stimulating', or 'significant' . . . (1967, p.24)

As we will see, the accounts of science given by the philosophers Kuhn and Feyerabend (Chapters 4 and 6), which challenge the more conventional views describd above, are based on versions of the coherence theory.

The pragmatic utility theory maintains that the categories 'true' and 'false' are irrelevant in judging the acceptability of theories. According to this view, whether a particular scientific theory is judged as acceptable or not depends only on whether it is useful or successful. Theories are not regarded as representations of reality as it is supposed to be, but rather as a means for calculation and prediction. Theories do not explain, they are merely convenient tools which can be used to manipulate the world, and if the application of a theory appears successful the theory is judged as acceptable. This view of truth seems to be implied by Osiander, writing in the preface to Copernicus's *Revolutions of the Hevenly Spheres*:

It is the duty of the astronomer to compose the history of the celestial motions through careful and skillful observation. Then turning to the causes of these motions or hypotheses about them, he must conceive and devise, since he cannot in any way attain true causes, such hypotheses as, being assumed, enable the motions to be calculated correctly from the principles of geometry, for the future as well as the past. The present author Copernicus has performed both these duties excellently. For these hypotheses need not be true nor even probable; if they provide a calculus consistent with the observations that alone is sufficient. (see Rosen 1959, pp.24–25)

The pragmatic utility theory is sometimes also known as *instrumentalism*. The idea of theories as instruments conveys very clearly what is entailed by this view, and is illustrated very well by the extract given above. Although the view is not associated with any particular account of the method of sciences, the instrumentalist attitude is one which is common amongst scientists. We shall meet it in the second part of this book which deals with such topics as modelling, measuring and experimenting.

Those who maintain the coherence or instrumentalist theories may believe, as do those who propose the correspondence theory, that there is an external reality. Thus they could all be described as realists. However, since the coherence and instrumentalist theories deny that this reality is knowable in any absolute way, there is a sense that both views are detached from realism. In what follows we will reserve the term realist for those views which also imply the correspondence theory of truth. To a large extent, realist and rationalist accounts of science go together. In the end, however, labels are unimportant. What is significant is the way these different attitudes in scientific work affect our judgements about what is or is not an acceptable account of the world.

The issue

All around us people claim to know things. On the basis of their intuition and experience they attempt to explain to us why things are as they are. There are textbooks on aspects of physical geography which are full of such explanations. Yet in comparison to what can be known about the world we probably know very little, and the accounts which are offered are undeniably less than perfect. As scientists the issue which faces us concerns the way in which we would respond to a question such as 'How do you know?'. The study of scientific method involves looking at the ways in which such a question might be answered. Some believe that scientific knowledge has a rational foundation while others reject any such view. We will examine what rationalism implies and the case for and against it in the other chapters in Part 1. In Part 2 we will examine the practical implications of these ideas for the work of the natural scientist.

CHAPTER 2:
THE CLASSICAL VIEW

'Letting the data speak for themselves'

Introduction

In Chapter 1 we described two contrasting views of science. According to one view, science is supposed to be a highly logical, ordered activity which tries to understand the world as it is independently of ourselves. According to the other view, the statements of science are no more than propaganda and its insights no deeper than those of the poet or mystic. In this chapter we will begin to examine the first of these opposing views.

Those who regard science as an ordered, disciplined activity can be described as rationalists. In using such a term we do not imply that there is a single coherent rationalist view of science. In fact, a number of divergent and often conflicting ideas can be recognized under the broad heading of rationalism. This chapter will describe one very influential version of rationalism, referred to here as the 'classical view'. The way in which people have reacted to these ideas will be described in the chapters which follow.

The classical view: outline

Russell (1961) has suggested that founders of modern science had two talents which are not necessarily found together. The first is immense patience in observation. The second is great boldness in framing hypotheses. He adds:

> The second of these merits had belonged to the earliest Greek philosophers; the first existed to a considerable degree, in the later astronomers of antiquity . . . and no one in the Middle Ages possessed either. (p.514)

He suggests that modern science had its origins in the seventeenth century with the work of men, such as Copernicus, Kepler, Galileo and Newton, who possessed both talents to a remarkable degree.

If we accept Russell's suggestion that modern science is to be characterized by the combined skills of imagination and observation we might ask a number of questions. How are these skills actually applied? How are the theories and laws which science produces to be justified? How are the observations and explanations of science to be judged?

One of the first systematic attempts to describe the methods of science was made by Francis Bacon in the early seventeenth century. Although his ideas have been qualified and refined, the tradition that he founded has persisted to the present day. It is this tradition that is the core of the classical view.

The ideas of Bacon were important because they questioned the view of science which had been handed down from Aristotle and which persisted until the seventeenth century. Aristotle regarded scientific reasoning as deductive, similar in character to the covering law model described in Chapter 1. However, although explanations were to be deduced from a set of general laws, these laws were not based on experience or observation in any systematic way but, rather, were obtained by insight. Bacon rejected intuition and insight as a basis for explanation. Instead he argued for the method of careful observation and experimentation. Laws and general axioms were still to be used as the basis of explanation but they were to be justified by experience rather than reason alone.

Those people who describe scientific method in terms of the classical view do not usually dispute the deductive model of explanation described in Chapter 1. Their attention tends to be focused on the way in which laws and theories are developed and justified. According to the classical model, scientific knowledge (theories, laws, explanations, etc.) is secure because it is obtained from experience. The classical account of scientific method is about the way in which experience is obtained and used.

A key element in Bacon's account of the scientific method is the concept of *induction*. In its broadest sense, induction is the process by which reliable generalizations are obtained from a set of observations of reality. According to the notion of induction generalizations are made once all the facts on a matter have been assembled. It is only by approaching matters in this way that the scientist can gain the necessary experience that will finally establish the truth of theories and hence the validity of explanations. Building on the idea of induction the classical view of scientific method can be summarized as follows:

1. the observation and recording of facts
2. the ordering and classification of these facts
3. the derivation of generalizations from the facts by induction
4. hypothesis formulation
5. attempted verification
6. proof or disproof
7. knowledge (theories, laws, explanations, etc.)

Knowledge so obtained is thought to be especially secure, because the first two steps are not assumed to make use of any guesses or hypotheses about how the observed facts might be interconnected. Bias is therefore avoided, and the objectivity of the results of scientific investigation assured. The claim that this allows for the greater objectivity of scientific knowledge compared to other kinds of knowledge is one of the main features of the classical view.

Some of the key assumptions of the classical view have been identified by Harré (1972, p.42). He recognizes three main principles:

1. *The principle of accumulation* which suggests that scientific knowledge consists of the conjunction of well-established facts, and that knowledge grows by the gradual accumulation of further well-attested facts.
2. *The principle of induction* which suggests that there is a form of reasoning which enables true laws to be obtained from a set of factual observations.
3. *The principle of instance confirmation* which asserts that the plausibility of a law is proportional to the number of instances which have been observed to conform to the law. The principle asserts that it is only through repeated observations that generalizations about the world are to be obtained and justified. The use of repetition is justified by the claim that nature is uniform.

These principles, Harré suggests, describe a 'very seductive theory of science'. To what extent can these methods be followed in the natural sciences?

The classical model and scientific practice

The classical model has been very influential in shaping people's view of scientific method. In his introduction to the *Origin of Species*, for example, Darwin writes that, after becoming interested in the problem of evolution,

> ... it occurred to me, in 1837, that something might perhaps be made out of this question by patiently accumulating and reflecting on all sorts of facts which could possibly have a bearing on it. After five years' work I allowed myself to speculate on the subject, and drew up some short notes ... (1958, p.27)

The influence of the classical view continues to the present day. For example, Chorley (1978) in his review of the bases of theory in geomorphology has identified the influence of 'functional theory'. Such work, he suggests, is based on the thesis that 'real world phenomena can be explained by showing them to be instances of repeated and predictable regularities...' Theory of this type, he adds, 'derives from the view that science is empirically based, rational, objective and aimed at providing explanation and prediction on the basis of observed and regular relationships' (p.2).

In order to get familiar with some of the ideas contained in the classical model consider the following case studies. The material of the case studies will be discussed in the sections which follow. You might also try Exercise 2.1.

Case Study 2.1: Variation of land plant numbers in the Galapagos

A major concern in biogeography has been to explain what controls the number of distinct species which can live on oceanic islands. Hamilton et al. (1963) suggest that the factors likely to regulate numbers may be divided into two broad groups. First, there are those which promote or impede interisland dispersal, such as the isolation of the island. Secondly there are those, such as island area or elevation, which control

Figure 2.1 The main islands of the Galapagos Archipelago

the number of opportunities or niches available for the species on each island.

In order to assess the importance of the different factors likely to control species richness, Hamilton et al. (1963) examined data for plant species number on 17 islands of the Galapagos group, see Figure 2.1. These data are shown in Table 2.1.

Analyse the data of Figure 2.1 and Table 2.1. What generalizations can you make about the factors controlling the number of plant species occurring on islands of the Galapagos group?

Explain exactly how you have arrived at your generalizations about the patterns shown by the data.

How would you go about trying to verify the generalizations that you have suggested?

On the basis of your investigation can you suggest any explanation of what it is that controls island species numbers in general, rather than plant species numbers in particular?

Island	Area (ml²)	Elevation (ft)	Isolation (mi)		Area of adjacent island (ml²)	Observed species numbers
	X₁	X₂	X₃	X₄	X₅	Y
1. Culpepper	0.9	650	21.7	162	1.8	7
2. Wenman	1.8	830	21.7	139	0.9	14
3. Tower	4.4	210	31.1	58	45.0	22
4. Jervis	1.9	700	4.4	15	203.9	42
5. Bindloe	45.0	1125	14.3	54	20.0	47
6. Barrington	7.5	899	10.9	10	389.0	48
7. Gardiner	0.2	300	1.0	55	18.0	48
8. Seymour	1.0	500	0.5	1	389.0	52
9. Hood	18.0	650	30.1	55	0.2	79
10. Narborough	245.0	4902	3.0	59	2249.0	80
11. Duncan	7.1	1502	6.4	6	389.0	103
12. Abingdon	20.0	2500	14.3	75	45.0	119
13. Indefatigable	389.0	2835	0.5	0	1.0	193
14. James	203.0	2900	4.4	12	1.9	224
15. Chatham	195.0	2490	28.6	42	7.5	306
16. Charles	64.0	2100	31.1	31	389.0	319
17. Albemarle	2249.0	5600	3.0	17	245.0	325

Source: after Hamilton et al. 1963.

Table 2.1 Insular number of land species and some other environmental factors for the Galapagos Archipelago. The numbers of the islands correspond to those in Figure 2.1. (X_3 = distance to centre of archipelago, X_4 = distance to nearest island)

Case Study 2.2: The history of the vegetation on the English Chalk
It has been suggested (e.g. Wooldridge and Linton 1933) that one of the factors responsible for the abundant Neolithic settlement on the Chalk Downs of southern England was the relatively thin tree cover which had developed in these areas so that man found them relatively easy to clear for agriculture. In order to examine this idea Thorley (1981) suggests that it is necessary to find out whether or not the Downs were ever well-wooded in the early post-glacial period prior to extensive occupation by man.

The reconstruction of the history of vegetation is not easy, although investigations can be made by the study of the pollen record contained in peat or lake deposits (see Moore and Webb 1978). As peat or sediment accumulates over time, a sample of the 'pollen rain' from the surrounding vegetation is deposited on it. Once the pollen is incorporated it is generally preserved because decomposition processes are inhibited by the waterlogged conditions. Thus if a core is taken through a peat deposit the fossil pollen can be extracted and identified at each level. Since peat or sediment accumulates by the addition of material at the surface, successively deeper levels within the deposit provide records of the fossil pollen further back in time. A picture of the fluctuations in vegetation which have occurred during the period of peat growth can then be built up.

The history of the vegetation recorded in a peat or sediment core can be summarized by means of a pollen diagram. Thorley (1981) presents two such diagrams for a site in the Vale of Brooks, East Sussex, one of which is reproduced in Figure 2.2. A radiocarbon date from a sample of the peat at 950–980 cm below the surface gave an age of 6290 ± 180 years BP (4340 ± BC).

Consider Thorley's suggestion that one needs to find out whether or not the Downs were well-wooded in early post-glacial times in order to test this idea that the area was relatively easy to clear. What generalizations can be made on the basis of the pollen evidence contained in Figure 2.2?

What other kinds of study would you make in order to verify your generalizations?

Verification and induction

Although many of the essential elements of the classical model can be traced to Bacon, further elements are derived from later writers. Particularly important revisions occurred in the nineteenth and twentieth centuries with the development of the ideas of *positivism*.

In its most general sense the term positivism describes the idea that enquiry and belief should be restricted to 'that which can be firmly established'. It is, therefore, a view which is similar to that of Bacon. In the nineteenth century Compte argued that not only was science based upon the ideas of positivism, but also the method of science (i.e. the acquisition of knowledge through experience) had wider relevance. Compte suggested, for example, that the methods of the sciences could be applied to the study of human affairs.

The influence of Compte in the social sciences cannot be considered here. The subject is explored more fully by Keat and Urry (1975), Gregory (1978) and Johnston (1983). For our purposes it is sufficient to note that the logical positivists in the twentieth century accepted the view embodied in the classical tradition that knowledge should be built on experience, and sought to give a more rigorous justification of this approach.

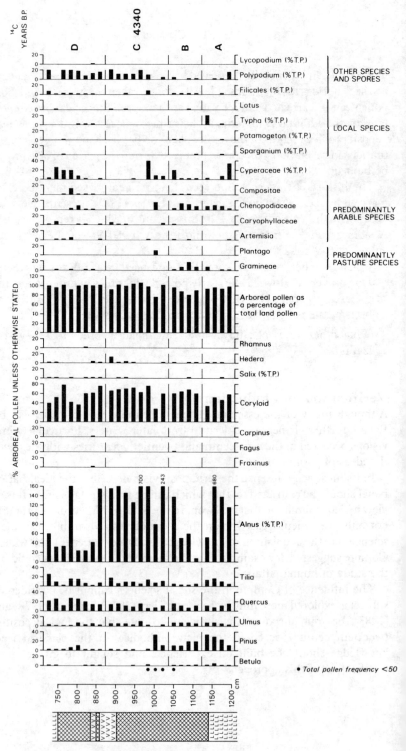

Figure 2.2 The pollen record from the Lewis I core
Source: after Thorley 1981

The logical positivists held that only two types of statement are significant or meaningful. The first consists of the statements of mathematics and logic which were held to be true by virtue of their form (tautologies). The second type of meaningful statements are those that concern matters of fact, and which were verifiable by observation. It is with this second class of statement that science is supposed to be mainly concerned. The positivists maintained that it was only because such statements of fact could ultimately be verified that scientific argument had any meaning.

The idea of verifiability was a key element of logical positivism and was enshrined in the so-called 'verification principle'. Ayer (1964) has stated the principle as follows:

> . . . a sentence is factually significant to any given person, if and only if, he knows how to verify the proposition which it purports to express – that is, if he knows what observations would lead him, under certain conditions, to accept the proposition as true, or reject it as being false. (p.35)

On the basis of this principle scientific statements were meaningful by virtue of their empirical basis. The principle describes the foundations of science and shows how it differs from other attempts to acquire knowledge. It provided what could be regarded as a criterion of *demarcation* showing what is and what is not genuinely part of science. According to the logical positivists, the experience provided by observation and experimentation was the thing which distinguished science from other pursuits as a means of acquiring knowledge. Activities which are not part of science were regarded as meaningless. The classical tradition asserts that knowledge grows by the accumulation of well-attested facts. Given such an assumption, the significance of the verification principle for the classical model should be clear. The idea of verifiability is important because it describes just how this firm factual basis is obtained. Having set up such a firm observational basis the way is open for the operation of the inductive method.

Limitations of the classical view

Despite its attractions the classical view is seriously deficient as an account of the methods of science. It provides neither an adequate description of the way in which science operates, nor a reasonable account of the way in which judgements are formed about scientific work. The limitations of the classical model arise in relation to three main problems, namely:

1. the problem of verification
2. the problem of induction
3. the problem of the relationship between observation and theory

In order to explore the character of the classical view of science more fully we will examine each of these problems in turn.

1. The problem of verification

According to the classical view scientific knowledge (theories, laws, explanations etc.) grows by the gradual accumulation of well-attested facts and by the operation of induction. Thus a key assumption is that a firm observational basis can be established so that inductive reasoning can be applied. The problem of verification is simply that no such firm basis can be achieved.

There are two aspects to the problem of verification. On the one hand there are the philosophical issues, such as: What is the status of the verification principle itself? Is it a tautology, that is, true by virtue of its logical form, or does it result from observation? Logical positivists held that there are only two kinds of meaningful statement, tautologies and those which are empirically verifiable by observation. Since the verification principle does not appear to fall into either category, by its own criterion it is meaningless. More importantly for the scientist, however, there are the practical concerns. Even if the verification principle can be shown to be 'meaningful', it would be necessary to show just how the truth of factual observations can be established.

The practical problem of verification is that there is a gap between the reality which is experienced and an interpretation of it described by an observational report. The scientist may hope that his observations correspond to reality, but there is no method to ensure that this correspondence is ever achieved. The scientist can never be certain that his senses are trustworthy and unprejudiced. Verification, in the sense of establishing the truth of a factual statement, is simply not possible. In the eighteenth century, Hume described the problem of verification as follows:

> If I ask [you] why you believe any particular matter of fact ... you must tell me some reason; and this reason will be some other fact, connected with it. But as you cannot proceed after this manner *in infinitum*, you must at last terminate in some fact, which is present to your memory or senses; or must allow that your belief is entirely without foundation. (Selby-Bigge 1902, p.46)

Ayer (1950), once the leading British exponent of logical positivism, eventually rejected the view he had held in his *Language, Truth and Logic*, that there are basic propositions (observations) which can act as ultimate verifiers. He grants that it is always possible that one can be mistaken.

Why should such arguments bother the practising scientist?

The problem of verification is important because it concerns the view taken of the 'raw data' on which the whole of the scientific process is supposed to be based. This point can be illustrated by reference to the material of Case Studies 2.1 and 2.2

The observational data describing the variation in plant species number, island area and the other environmental data for the Galapagos may seem fairly straightforward and reliable. Yet, there is no final means of establishing whether they faithfully represent the situation on these islands. The data can only be justified by reference to other factual statements, such as 'the same census method was used on each island' or 'no plant species not included in the census can now be found on the islands', but such statements are no more secure than the original observations contained in Table 2.1. In the end we take such data on trust unless anything can be found to undermine our confidence in them.

The speculative nature of the observational basis of science is illustrated even more clearly in Case Study 2.2. In contrast to the straightforward measurements of island area and isolation, the measurements which formed the basis of the pollen record are much more open to doubt. The assumption of their reliability depends on a number of beliefs, including the idea that the pollen rain is representative of the surrounding flora, and that there is no selective preservation once the pollen is incorporated into the peat deposit.

In selecting the papers of Hamilton et al. (1963) and Thorley (1981) as the basis for our discussion we do not mean to imply that their factual basis is especially suspect. Critics of the classical tradition would argue that any scientific paper must face the same kind of problem (see Medawar 1963; Edwards 1983). The truth of propositions, even of the most simple kind, is never beyond doubt. If this is so then it is difficult to see how the inductive method guarantees the truth of the generalizations which it is supposed to generate.

2. The problem of induction

Let us assume that, despite the problem of establishing the truth of the set of observations, it is at least possible to achieve a very high degre of confidence in the empirical reports that are used in the inductive process. Such confidence might come from the fact that we can find no grounds to fault the observations. Would it then be possible to proceed according to the classical tradition? The critics of the classical view argue that it is still not possible to work in this way. In support of their claim they would point to the logical problems associated with the idea of induction itself.

As we have seen the inductive method tries to argue as follows:

All observed a's are b's
Thus all a's are b's   (1)

where, for example, the a's and b's might be:

All observed mammals are terrestrial
Thus all mammals are terrestrial

or:

All observed precipitation events observed in Nottinghamshire were less than 15 cm/hr.
Thus all storm events will be less than 15 cm/hr.

The problem of induction was stated clearly by the philosopher David Hume in the eighteenth century. He asked:

Are we justified in reasoning from repeated instances of which we have had experience to other instances or conclusions of which we have none? (see Popper 1972b, p.4)

His answer was 'no'. There is no principle that can justify the truth of a conclusion derived from a set of statements about particular events. The conclusion of the inductive argument does not follow 'of necessity' in the same way as, for example, the conclusion of a deductive argument does (see Chapter 1). This is so regardless of how many observations are made, and despite the fact that the observations themselves might be true reports about the world. The problem of induction is simply that *there is no logical principle of induction*. This conclusion is important because it completely undermines the classical view.

Despite the lack of any logical principle of induction, some people have tried to defend it as a method of reasoning. Two kinds of defensive argument have been tried. The first introduces the idea of probability. The second appeals to experience. Let us examine each of these ideas in turn.

Although the conclusion of an inductive argument does not follow as a matter of logical necessity, supporters of the classical view often suggest that since a large number of repeated observations have been made the conclusion must be *more probable*. In order to include this idea about degree of certainty the simple model of an inductive argument shown in (1) could be rewritten as follows:

All observed a's are b's
Thus all a's are probably b's.   (2)

Unfortunately this form of reasoning encounters the same problem as the initial scheme. One is still seeking to justify a universal statement on the basis of a set of particulars. Such a tactic has no logical basis because the conclusion does not necessarily follow from the observations. In addition, the idea of the probability of a hypothesis being true is itself highly misleading. It is possible to determine the relative probability of an event in a sequence of events. This is simply the relative frequency with which that event occurs. Similarly it is possible to calculate the probability of a statement in a sequence of statements (say a sequence of statements describing a series of events). However, one cannot calculate the probability of a hypothesis being true unless one can specify with reference to what sequence of other hypotheses the probability is to be assessed. Is the probability to be calculated as the frequency of observation statements which either agree or disagree with the hypothesis? Or is it to be calculated on the basis of the frequency of tests which the hypothesis has passed in relation to the sequence of tests undertaken or which might be undertaken? The probability that a hypothesis is true cannot be calculated in any such way.

Popper (1972a, p.255) suggests that the idea of the probability of the truth of a hypothesis arises through a confusion of psychological and logical questions. As the result of experience one may build up different degrees of confidence about different hypotheses, but such psychological beliefs do not form the basis of any rational scientific method. It is not possible to translate such subjective feelings into statements of probability about a hypothesis being true.

The arguments about the probability of a hypothesis being true are difficult but they are nevertheless very important because questions of probability frequently arise in scientific work. In order to clarify the debate it will be valuable to reconsider the inductive pattern of reasoning shown in (2), above. It should be noted that although the model includes the idea of probability it does not deal with statements which are themselves probability statements. In order to see this difference consider the following:

Of all the a's observed a proportion, p, are b's
Of all a's a proportion, p, are b's   (3)

Although it is valid, on the basis of a sample, to ascribe a certain confidence to the estimate of the proportion p, this is not the same as calculating the probability of the truth of the hypothesis that the proportion is indeed p. The confidence limit for p describes the range within which we expect its true value to lie, not anything like a probability that a particular hypothesis about the value of p is true. Although *statistical inference* is a valid method of analysis, it is not a means of establishing the truth of any particular hypothesis in the manner implied by the classical tradition.

Finding that the appeal to probability is not valid, supporters of the inductive method often turn to 'experience' as an alternative means of justifying the approach. Inductive arguments, it is often claimed, have been found to work in the past. Unfortunately, even this strategy is unsuccessful because the appeal to experience would have to be based on something like the following argument:

Induction worked successfully on all observed occasions
Thus induction works on all occasions.

The argument is clearly suspect. The attempt to appeal to experience uses the same kind of inductive argument that one is trying to justify. Moreover, if we are prepared to base our acceptance of induction on experience, what would we say the first time application of induction yields a false hypothesis?

The problem of induction is a major obstacle for the classical view. Despite such difficulties scientists have continued to operate as if induction is an acceptable way of working. Unfortunately, the tendency has been reinforced with the widespread use of statistical methods. This can be illustrated by returning to the problem which Hamilton et al. (1963) faced in evaluating the results of their analysis (see Case Study 2.1).

Hamilton et al. (1963) employed simple and multiple regression techniques as a means of describing the relationship between the independent variables, area, isolation, etc., and the dependent variable, plant species number. Both linear and curvilinear regression models were employed. The results are summarized in Table 2.2.

In Table 2.2 the coefficient of determination is a measure of how well an independent variable, or group of variables, explains the variation in the dependent variable. The single variable which gave the highest coefficient of determination for species number was island area when the curvilinear model was employed. The amount of

Table 2.2 Factors controlling plant species number on the Galapagos

Model	Independent variables					R^2
	X_1	X_2	X_3	X_4	X_5	
Simple						
linear		*				0.45
curvilinear	*					0.67
Multiple						
linear		*	*	*	*	0.84
curvilinear	*					0.67

Variables defined as in Table 2.1. * indicates a significant coefficient.
R^2 is the coefficient of determination which expresses the amount of variation in the dependent variable (island species richness) explained by the independent factor(s).
Source: after Hamilton et al. 1963

variation accounted for by the independent variables was higher, however, when multiple regression was used. The coefficient of determination was highest for the multiple linear model. With this model, island elevation and the measure of island isolation were the most important variables.

Since several models explained a significant proportion of the variation in species number, Hamilton and his co-workers had to make a choice between them. The different models suggested that different independent variables were significant and they could not all be true. How did these workers make the choice? In their evaluation of the different models Hamiliton et al. (1963) argue: 'It is not that one model predicts or fails to predict, but that several predict with varying degrees of accuracy . . .' and as a result they suggest: '. . . the model which estimates most precisely the primary measurements . . . is more likely to quantitatively represent organismic responses to environmental variants' (p.1576). In other words the model which gives the best prediction is to be accepted. As a result Hamilton and his co-workers conclude that in areas such as the Galapagos, where a large number of endemic species have evolved, ecological diversity (represented by island elevation) and isolation are the key factors controlling island species richness.

Although methods of statistical analysis, such as regression, are not a form of inductive reasoning Hamilton uses them as if they were. However, the awkward fact is that their application leads to conflicting conclusions. How can one decide between them? The approach employed by Hamilton et al. (1963) is highly suspect. On the one hand they suggest that correlation necessarily implies causation, an assumption that is itself questionable. On the other, they imply that predictive success is the criterion by which the acceptability of any particular statistical generalization should be judged. This kind of argument is a form of instrumentalism (see Chapter 1).

In forming an assessment of the study by Hamilton et al. (1963) it is valuable to consider the subsequent work by Johnson and Raven (1973). Their study also employed regression techniques but this time applied them to data from a new flora of the Galapagos produced by Wiggins and Porter (1971). Johnson and Raven found a multiple curvlinear model to have the highest coefficient of determination, with island area making the only significant contribution. It is seen that similar techniques applied to similar data generated yet more conflicting conclusions.

Quite apart from the fact that multiple regression techniques may not be appropriate for the analysis of the kinds of data on which the study of species numbers is based (see Vincent 1981), the attempt to use statistical techniques in this inductive manner is clearly fraught with difficulties. Statistical techniques can no more guarantee the reliability of a generalization than the more qualitative analysis employed, for example, in the study of the history of the vegetation of the English Chalk (see Case Study 2.2). The idea that induction can provide a rational basis for reasoning is simply misleading. Any number of consistent interpretations can be devised using such data. None have any prior claim to validity.

3. The problem of theory-dependent observations

A particularly seductive feature of the classical approach is its claim to objectivity. According to the classical tradition, observations are made in an unbiased way without any theoretical assumptions. Hypotheses and theories emerge much later, after a careful investigation of the data. In applying this approach there is a sense in which the scientist is 'letting the data speak for themselves' (cf. Gould 1981). Unfortunately, the division between theory and observation is by no means as simple as the classical view suggests.

As we saw in Chapter 1, an observation is a singular statement describing an event or circumstance at a particular time or place. In Case Study 2.2, for example, the statement that

> The sample of well-humified peat from the 950–980 cm level of the Lewis I core had a radiocarbon age of 6290 ± 180 BP

would conform to such an idea. In contrast, a theoretical statement might have the following form:

> Radiocarbon, ^{14}C, is produced by the bombardment of atmospheric nitrogen by cosmic rays.

This is a universal statement which contains reference to things which are not directly observable. In fact the statement can only be properly understood in the context of atomic theory.

The classical tradition suggests that observational and theoretical statements such as those illustrated above are fundamentally different in character. Observations are supposed to depend only on the experience of our senses, while theoretical state-

ments depend on our accumulated understanding. Moreover, the two types of statement have quite different functions in science. The classical view represents
observation statements as a kind of basic building block from which general theories
are constructed. Critics of the classical view deny this differentiation between theory
and observation. The problem of theory-dependent observations arises because the
distinction cannot be supported.

The problem of theory-dependence in observation arises in two ways. The first
concerns the individual terms employed in the observational statement and the
second the relationship of the observation to the programme of research as a whole.
We will consider each of these aspects of the problem in turn.

The critics of the classical view deny that observations can be made in an unbiased
way, free of theory, because theoretical terms *always* enter into observational reports.
Popper (1972a, p.94), for example, argues that every description uses *universal*
names (symbols or ideas) so that every statement has the character of theory. He
illustrates his point by considering the simple observation 'here is a glass of
water':

> The statement . . . cannot be verified by any observational experience. The reason
> is that *universals* which appear in it cannot be correlated by any specific sense
> experience . . . By the word 'glass', for example, we denote physical bodies which
> exhibit certain *law-like behaviour*, and the same holds for the word 'water'.
> (p.95)

In other words, if we were pressed to justify our reports that the container was glass
or the substance water, we would be forced to appeal to other statements like 'if this
substance were mixed with oil it would separate'. If pressed further we might add '. . .
since water and oil do not mix'. Finally, we might be forced to give an explanation in
terms of the molecular properties of liquids, and so on. The report that we see a glass
of water is not reducible to a single experience but depends on our belief that the substance in question belongs to the class of things with certain general properties which
we denote by the term 'water'.

The fact that observations depend on theories about the world is further illustrated
by the statement about the peat sample in Case Study 2.2. The description 'well-
humified peat' depends no more upon simple sense-experience than did the description of the substance 'water' in the example given above. The observations made on
the peat also illustrate the point that, quite apart from description, the act of
measurement is also theory-dependent. The concept of radiocarbon age, for example,
depends upon the acceptability of the underlying theory of radiocarbon dating. Some
simple features of such a theory are illustrated in Figure 2.3. The theory not only
makes asumptions about the mechanisms for the generation of ^{14}C, but also assumes
that the ratio of ^{14}C to ^{12}C has been constant over time, and that ^{14}C decays at a
certain rate (see, for example, Worsley 1981a).

The discussion of the problem of theory-dependent observations has focused upon
the level of the individual terms which make up the factual report. However, as noted

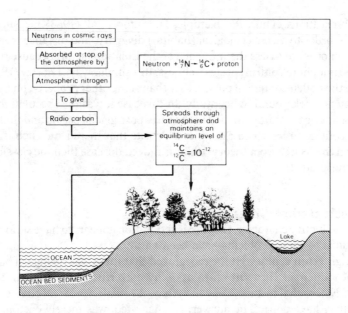

Figure 2.3 The basis of ^{14}C dating. The ocean and the biosphere are in equilibrium with the atmospheric pool of CO_2. When the organic material or carbonates contained in the oceans are incorporated into sediments there is no further CO_2 exchange. Since ^{14}C decays at a constant rate, the amount of activity lost can be used to estimate the age of the sediment.

above, the problem has a second aspect. This concerns the relationship of the reported observations to the research programme as a whole and further undermines the idea that observations can be made independently of theory.

The classical approach proposes that observation precedes theory formation. If this is so, then on what basis are the observational reports selected in a given programme of research? Let us consider this question in the context of Case Studies 2.1 and 2.2. We might ask: Why did Hamilton et al. (1963) select factors such as area and elevation as objects of measurement? Why did they not report on, say, island shape or climate? Similarly we might ask: Why did Thorley (1981) choose to observe the pollen content of the peat rather than, say, its calorific value?

Questions about the reasons for measuring particular things may seem naive, but in the context of the classical model they are extremely important. If, according to the classical model, observations are made prior to theory, then it is legitimate to ask on what basis the particular observations were selected. The observations were not made randomly without some thought as to what was required, but some judgement was exercised as to which observations were relevant to make and which were not. What supplied this criterion of relevance?

The critics of the inductive view argue that it is theory whch guides the scientist in making observations. Thus, for example, Hamilton et al. (1963) measured island elevation in order to assess ecological (habitat) diversity, and distance from nearest neighbour in order to assess the influence of the isolation factor, because these were the factors thought to control species richness. In a similar way, Thorley (1981) chose to observe the pollen content of peat, rather than some other property, on the basis of the theoretical background to her study. In this case it was the speculation that the chalk must have been thinly wooded in the early post-glacial period and thus easier to clear in Neolithic times. Such examples illustrate that there is no such thing as an observation that is free from theory. If this is indeed the case then the classical model is largely misguided.

The classical view: interim review

Despite its seductive character, the classical tradition appears to have serious defects. The reactions to these problems have been various.

In the eighteenth century Hume recognized that there was no logical basis to the method of induction. He went on to ask why, nevertheless, all reasonable people expect and *believe* that instances of which they have experience will conform to those of which they have none. The answer, he suggested, was merely custom or habit. Russell summarizes Hume's position very clearly:

> He starts out . . . with intentions of being sensible and empirical, taking nothing on trust, but seeking whatever instruction is to be gained from experience and observation. But . . . he arrives at a disastrous conclusion that from experience and observation nothing is to be learnt. There is no such thing as rational belief: 'If we believe that fire warms, or water refreshes, 'tis only because it costs us too much pains to think otherwise'. We cannot help believing, but no belief can be grounded in reason. (1961, p.645)

The problems of the classical tradition which we have described here imply that not only is its claim to represent a rational basis of acquiring knowledge fallacious, but, in any case, the method is impossible to operate. Its acceptance appears to be based on dogma rather than reason. The unworkable nature of the classical approach is in fact hinted at by Darwin, despite his acknowledgement that his work conformed to the 'canons of inductive reasoning'. He wrote in a private letter of 1861:

> About thirty years ago there was much talk that geologists ought only to observe and not theorise; and I well remember someone saying that at this rate a man might as well go into a gravel-pit and count pebbles and describe colours. How odd it is that anyone should not see that all observation must be for or against some view if it is to be of any service. (see Gruber 1974, p.122)

Elsewhere, in 1863, he wrote 'let theory guide your observations'. For a more extensive discussion see Gruber (1974) and Darwin (1903). Given the difficulties associated with the classical approach which such views highlight, we would suggest that neither of the papers on which Case Studies 2.1 and 2.2 are based are examples of the inductive method in operation because it is an unworkable method. There is no way in which the scientist can go into the field and simply 'observe'. The aims of science cannot be achieved by the inductive route and any attempt to conform to the classical model is an empty gesture. If work such as that described in Case Studies 2.1 and 2.2 has added to scientific knowledge, it is not because it employed inductive reasoning, even if the workers thought they followed the classical tradition. To what kind of logic (if any) do such studies conform?

To the practising scientist all this may seem no more than opinion. Science, it could be argued, is too complex an activity to be written off that easily. Hume's conclusion about induction is, however, of fundamental importance for any who maintain that the classical model does capture the essentials of scientific method. If the conclusion is valid then every attempt to arrive at general laws and theories by means of observations is flawed. If this is so the rationality of science is a sham. Russell reacted to Hume's conclusion as follows:

> It is . . . important to discover whether there is any answer to Hume within the framework of a philosophy that is wholly or mainly empirical. If not there is no intellectual difference between sanity and insanity. The lunatic who believes that he is a poached egg is to be condemned solely on the ground that he is in a minority, or rather – since we must not assume democracy – on the ground that the government does not agree with him. This is a desperate point of view, and it must be hoped that there is some way of escaping it. (1961, p.646).

A possible escape route will be explored in the next chapter.

CHAPTER 3:
THE CRITICAL RATIONALIST VIEW

'A woven web of guesses'

Introduction

How is one to escape from the conclusion that science is irrational? One solution has been offered by Karl Popper. His account of science is an important alternative to the classical tradition. In this chapter we will describe the view of science which Popper has provided.

Popper developed his ideas about science in response to the problem of induction, and to the views which the logical positivists held about verification (see, for example, Popper 1976, pp.78–90; 1983, pp.11–18). Popper rejects both induction as a method and also the belief that verifiability is the hallmark of scientific statements. Instead he argues that scientific method is essentially deductive in character, and that it is the ability to falsify scientific statements, rather than to verify them, which distinguishes scientific statements from all others.

Critical rationalism: outline

Popper begins by accepting Hume's conclusion that there is no principle of induction which permits us to derive universal laws from a set of observations. He agrees that generalizations cannot be verified, or be *proved to be true* by empirical evidence. Despite such conclusions, however, Popper continues to believe that it is possible to base judgement on observation, and to see science operating in a framework which is wholly or partly empirical. The basis of this conclusion is the realization that although there is no method whereby observations can be used to verify a universal law, potentially only one counter-observation is necessary to *refute or falsify* it. Let us look at this idea more closely.

Consider the (notorious) law-like, universal statement that:

All swans are white.

No number of sightings of white swans can establish the truth of the generalization that all swans are white. A single report of a black swan can, however, show the generalization to be false. Thus, while belief in the truth of a proposed law cannot be justified, such claims can be *tested* by observation. Although proof is not possible:

... the assumption of the truth of test statements sometimes allows us to justify the claim that an explanatory universal theory is false. (Popper 1972b, p.7)

One of the key ideas in Popper's account of scientific method is illustrated by this example. It is the logical proposition that there is an *asymmetry* between verification and falsification (Popper 1972a, pp.41–43; 1983, pp.131–139). That is, while universal statements cannot be verified they can be falsified. The recognition of this asymmetry is important, because not only did it allow Popper to resolve Hume's problem concerning the irrational basis of belief, but it also allowed him to provide an alternative to the classical view. Moreover, Popper was able to show just what it is that distinguishes science from all other activities.

Hume concluded that science was irrational because the inductive method had no logical basis and people nevertheless appeared to act as if repeated observations could justify belief in general laws. Popper accepts Hume's conclusion that there is no such thing as induction, and that the attempt to use it as a method of justifying belief is irrational, but goes on to show that a case can still be made for believing that science is rational. Popper was able to avoid the conclusion that science is irrational because he was able to identify another way of proceeding. The approach described by Popper is the method of proposing hypotheses and then exposing them to the severest criticism possible in order to discover whether they are false (see Popper 1976, p.86). Good reasons can be given for holding one theory rather than another, but such reasons are not provided by anything like verification. According to Popper, justification is provided by a process of critical investigation. Let us examine these ideas in more detail.

Popper considers that the aim of science is to solve problems (Popper 1972b, p.131; 1974, pp. 122–3; 1983, p.131). The covering law model described in Chapter 1 illustrates what such 'problem solving' represents. To solve a problem means to obtain an explanation of an event from a set of universal laws and a description of the particular circumstances associated with the event (the initial conditions). The scientist may be able to think up a number of theories in order to solve the problem. Popper (1972a, p.32) argues that a judgement can be made about the adequacy of these competing trial solutions on the basis of the results of critically testing the underlying theories. Such a process exploits the same deductive logic as that described in the covering law model.

Popper (1972b, p.119) summarizes his outline of scientific method as follows:

$$P_1 \rightarrow TT \rightarrow EE \rightarrow P_2$$

Faced with some initial question (P_1) the scientist attempts to resolve it by putting up some trial solution or tentative theory (TT). This trial solution may be a new idea or it may be derived from an existing body of knowledge represented by a theory. It is important to note, however, that the solution offered cannot be justified in any way. It is, Popper suggests, only an anticipation. The trial solution can, nevertheless, be tested. Using deductive logic, the scientist can predict what else must follow if the theory underlying the trial solution is true. These predictions can then be tested in an attempt to refute the trial solution if it is false. This is the stage of error elimination (EE). The assessment of the speculative theory in the light of the results of these tests, leads on to a new set of problems (P_2).

The notion of falsification provides Popper (1972a, pp.40–44) with a means of showing just what it is that distinguishes scientific statements from non-scientific ones. For the logical positivists the *demarcation criterion* was that statements were verifiable. As we have seen the idea was unworkable. As an alternative Popper suggests that statements are considered scientific only if they are potentially falsifiable. In relation to the model described above, it is the fact that the trial solution is testable, and that it can be refuted if it is false, which establishes its claim to be scientific.

Popper describes his view of science as *critical rationalism* (Popper 1974, p.26). The word critical comes from the fact that scientific method is essentially critical in character. The term rationalism is used because such critical investigation is supposed to provide good (rational) reasons for holding some theories rather than others. The key assumptions of the critical rationalist view of science may be summarized as follows:

1. *The principle of falsifiability* which states that universal statements and theories can only be refuted and not verified.

2. *The principle of criticism* which states that since all scientific knowledge is speculative the only rational attitude to adopt towards it is a critical one. Scientific knowledge grows by a process of trial and error, rather than the gradual accumulation of well-attested facts.

3. *The principle of demarcation* which states that the essential characteristic of scientific statements is that they are empirically testable, that is, capable of refutation if they are false. This characteristic of scientific statements is important, because it is only if they are potentially falsifiable that their correspondence to the truth can be tested by critical investigation.

The critical rationalist model has been summarized in Figure 3.1. Such summaries are not very helpful unless we try to find out what it feels like to apply such ideas.

The critical rationalism and scientific practice
In order to get familiar with the ideas of the critical rationalist, work through the following case studies. The material will be discussed in the sections which follow. You might also try Exercise 3.1.

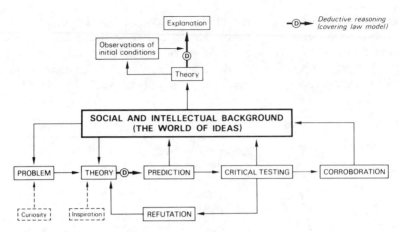

Figure 3.1 Critical rationalism

Case Study 3.1: The effect of island area on species richness

The work described here deals with the same problem as that considered in Case Study 2.1, namely: What factors control the number of species resident on islands? In contrast to the approach based on regression, Simberloff and Wilson (1970), and later Simberloff (1976), used experimental methods to test the theory that island species richness is directly controlled by island area.

According to the theory of island biogeography (see, for example, Williamson 1981), area could control island species richness in two ways:

1. *Directly:* by the influence of area on the probability of a population becoming extinct. The larger the area available for colonization the larger the potential size of the resulting species populations. In variable environments larger populations have a lower chance of extinction because their greater numbers provide a kind of buffer.
2. *Indirectly:* if area is simply a surrogate for environmental diversity. The larger the area of the island, the greater the number of distinct habitats and niches available, and the larger the number of species which can be supported.

The two possible effects of island area are quite distinct. One can imagine two islands of different size, but with the same number of habitat types or niches. The direct effect would operate if it was the area of each type which controlled the number of species through its effect upon population size. Clearly both direct and indirect effects could be superimposed upon each other and in practice it could be very difficult to isolate each factor. In order to cope with such problems, Simberloff and Wilson (1970) and Simberloff (1976) devised a set of large-scale experiments.

Simberloff and Wilson investigated the number of arthropod insects which inhabit mangrove swamp islands in the Florida Keys. Since the islands consisted *only* of mangrove, they considered that the number of niches or habitat types on each island would be independent of the size of the islands. They investigated the direct effect of island area by experimentally manipulating species richness and island size.

In the first set of experiments (Simberloff and Wilson 1970) islands were fumigated and all insect species were removed. Colonization was then allowed to take place, and

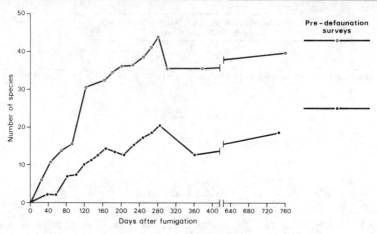

Figure 3.2 The pattern of colonization by arthropods on two mangrove islands in the Florida Keys following complete fumigation
Source: after Simberloff 1976

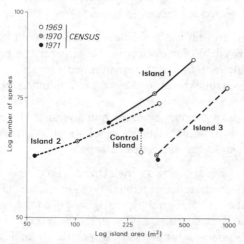

Figure 3.3 The effect on the number of arthropod species of reducing the size of mangrove islands. Islands 1 and 2 were reduced after the 1969 and 1970 census. Island 3 was reduced in area only once, after the 1969 census. The size of the control island was not changed.
Source: after Simberloff 1976

surveys were conducted in order to find out how many species returned and whether the same number of species recolonized the islands. The results of this experiment are shown in Figure 3.2.

In the second experiment (Simberloff 1976), sections of several mangrove islands were removed down to water level by teams of workers using power saws. A census of the species present on each island was made over the three years following the reduction of island area. The results of this experiment are shown in Figure 3.3.

In Figure 3.2 what do you think is the purpose of including the data on the number of species present before fumigation?

In Figure 3.3 what do you think is the purpose of including data for the control island? This is the only one of those studied whose area was not changed.

What do you consider the implications of the results to be in the light of the original problem? Try to summarize the steps of the argument using Popper's summary of scientific method.

Case Study 3.2: The 1750 end-moraine hypothesis

In reconstructing the post-glacial history of glacier movement in southern Norway, it has been suggested that a major advance occurred during the mid-eighteenth century. The cold period represented by this advance is sometimes described as the Little Ice Age.

The advance represented by the Little Ice Age is considered to have been of sufficient magnitude to have allowed glaciers to reoccupy their valleys and to have destroyed any moraines which had formed earlier. Thus the outermost, or end-moraines in the sequences now found around southern Norwegian glaciers are often attributed to the Little Ice Age advance. The date of this advance is usually put at about 1750. This date is largely based on the historical evidence from the glacial foreland at Nigardsbreen (see Andersen and Sollid 1971; Ostrem et al. 1976). Contemporary tax returns show that an advancing glacier destroyed a farm in the valley prior to 1748. It is assumed that many outer-moraines in southern Norway probably date from this time. Unfortunately, historical records that would enable geomorphologists to date moraines at each glacial foreland in southern Norway are rare. Hence indirect methods of dating must be used. Several techniques are available, including lichenometric dating, and radiocarbon dating of fossil soils (palaeosols) and organic deposits buried under moraines.

As bare rock surfaces are exposed by retreating glaciers they become available for lichen colonization. Lichenometric dating (see Locke et al. 1979; Worsley 1981b) is based on the assumption that the size of a lichen found on an exposed glacial deposit depends on the time which has elapsed since colonization. If a representative measure of lichen size can be obtained for moraines of known age, then dates can be interpolated for moraines whose age is unknown (see Figure 3.4). Matthews (1974, 1975) used this technique to date the moraines which occur below Storbreen in southern Norway (see Figure 3.5 a and b).

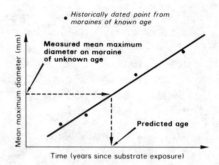

Figure 3.4 The basis of lichenometric dating. A dating curve is constructed using the mean maximum diameters of lichens on moraines of known age. The age of a moraine whose age is not known can then be predicted by measuring the mean maximum lichen size on the deposit.

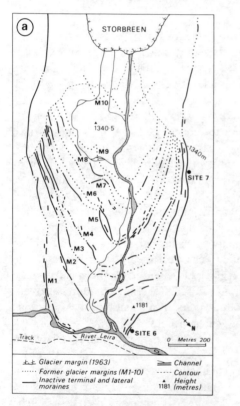
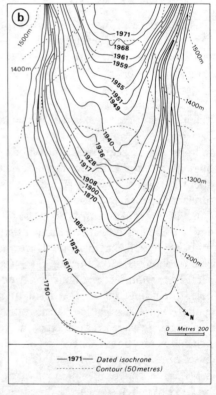

Figure 3.5 (a) The 10 former glacier margins at Storbreen, southern Norway; (b) the complete areal chronology for Storbreen based on lichenometry. (Sites 6 and 7 relate to Figures 3.7 and 3.8, and Table 3.1.)
Source: after Matthews 1974, 1975 and 1977

In dating the moraines at Storbreen, Matthews (1974, 1975) calibrated his lichenometric curve by assuming that the outermost moraine dated from 1750. No historical data were available, except that from Nigardsbreen. In order to test this assumption he set up a series of lichenometric curves using only those points for which historical evidence was available, and then *extrapolated* the curve to predict dates for the outermost moraine (Matthews 1977). The curves were based on the single largest, the mean of the five largest, and the mean of the ten largest lichens per moraine. The results of his analysis are shown in Figure 3.6. Using his dating curves Matthews was able to predict the mean age and the 95% confidence limits for the estimate for the outermost moraine.

What do you consider to be the status of the 1750 end-moraine hypothesis in the light of Matthews's (1977) results?

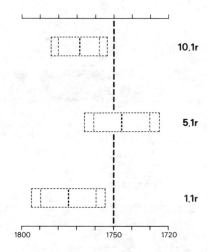

Figure 3.6 Mean age of an end-moraine at Storbreen, southern Norway predicted using lichenometric dating. The estimates were made using dating curves based on the single largest, and the means of the five and ten largest lichens per sample area. The 95% and 99% confidence limits to the mean estimate are shown.
Source: after Matthews 1977

As a further test of the 1750 end-moraine hypothesis, Griffey and Matthews (1978) excavated the deposit buried beneath the outermost moraines at several glacial forelands, including the one at Storbreen (see Figure 3.7). The radiocarbon dates obtained for samples at their study sites are shown in Table 3.1 and Figure 3.8. In making their observations Griffey and Matthews suggest that the samples taken for dating:

> . . . are considered to represent buried soil surfaces and the resulting ^{14}C dates only require correction for the mean residence time, and short-term variations in atmospheric ^{14}C, to become estimates of moraine ridge age. (p.83)

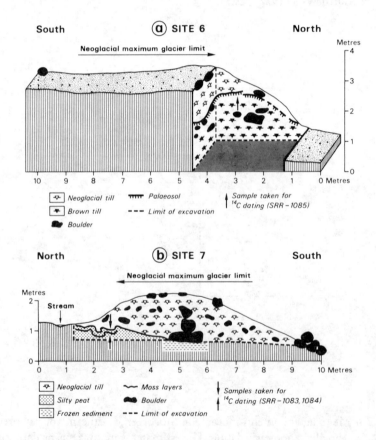

Figure 3.7 Cross-sections through the end-moraine at Storbreen, southern Norway showing location of samples used in radiocarbon dating. The location of the sites is shown in Figure 3.5a.
Source: after Griffey and Matthews 1978

Table 3.1 ^{14}C dates from palaeosols and moss layers associated with some outermost Neoglacial moraine ridges, Jotunheimen

Glacier	Site No.	Sample type	Fraction dated	Lab. No.	^{14}C years	δ^{13} C
Styggdalsbreen	2	Palaeosol (top 2.5 cm)	Fulvic acids	Birm-720	A 2770±200	−18.23%
			Humic acids		B 2830±130	−21.82%
	2	Palaeosol (top 2.5 cm)	Fulvic acids	Birm-805	A 1930±120	−22.77%
			Humic acids		B 2740±100	−22.30%
Lierbreen	4	Palaeosol (top 2.5 cm)	Fulvic acids	Birm-721	A 1200±260	−18.56%
			Humic acids		B 1490±120	−21.56%
	5	Palaeosol (top 2.5 cm)	Rootlets	Birm-806	A 1310±100	−22.68%
			Residual		B 1900±100	−23.52%
			Humic acids		C 1740±100	−28.08%
Storbreen	6	Palaeosol (top 2.5 cm)	Complete sample	SRR-1085	1070± 40	−27.1%
	7	Upper *Sphagnum* layer	Complete sample	SRR-1083	644± 45	−24.4%
	7	Lower *Sphagnum* layer	Complete sample	SRR-1084	532± 40	−25.4%

Source: after Griffey and Matthews, 1978.

Thus the deposition of the moraine is assumed to have stopped the development of the buried soil and peat deposits so that their radiocarbon age reflects the length of the time period since the moraines were formed.

Do the radiocarbon dates obtained by Griffey and Matthews change your view of the 1750 end-moraine hypothesis?

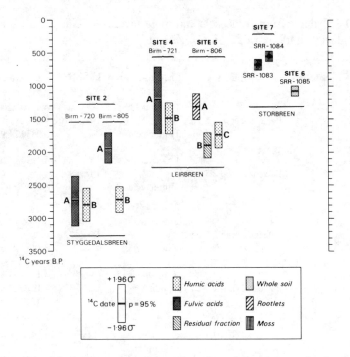

Figure 3.8 Graphical representation of radiocarbon dates obtained for end-moraine deposits in southern Norway.
Source: after Griffey and Matthews 1978

Case Study 3.3: The origin of the crater at Coon Butte, Arizona
Gilbert (1896) investigated the acceptability of several different hypotheses put forward to explain the origin of the crater at Coon Butte, Arizona (Figure 3.9 a, b and c). The crater is to be found in the arid north-east corner of Arizona. Gilbert describes the crater as several thousand feet wide and a few hundred feet deep, standing on a plain underlain by limestone. Many fragments of iron are found on the slopes of the crater and on the surrounding plain. The rim of the crater appears to be composed of limestones and sandstones.

Gilbert considered the various hypotheses which had been put forward to explain the origin of the crater. The main hypotheses were:

1. *Explosion crater:* the crater resulted from some kind of explosion which threw the iron fragments out across the surrounding plain (no cause of the explosion was specified).
2. *Meteorite impact:* a shower of falling meteorites included one larger than the rest, and its greater mass, 'by the violence of its collision, produced the crater' (Gilbert 1896, p.4).
3. *Volcanic intrusion:* although the rocks of the rim are not volcanic, the inclined

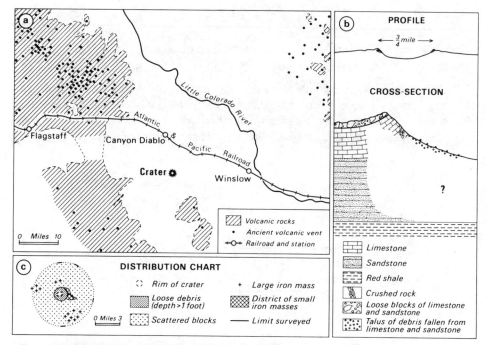

Figure 3.9 The crater at Coon Butte, Arizona: (a) location; (b) cross-section through crater; (c) map of deposits around the crater
Source: after Gilbert 1896

strata overlay a volcanic intrusion (a laccolith). This intrusion caused the strata to arch and central subsidence to occur, thus forming the crater.
4. *Volcanic explosion:* a body of steam produced at depth by volcanic heat caused an explosion at the surface which formed the crater.

In the case of explanations 3 and 4, the presence of the iron was regarded as coincidental and not connected with the origin of the crater.

Consider each of these hypotheses in turn. Try to suggest some observations which could be made in order to test them. You might be able to examine some of your deductions by looking at Figure 3.9.

Case Study 3.4: Witch doctors and rainfall

Sprout (1984) examined one possible cause of rainfall, namely the effect of witch doctors dancing. He set up the following experiment:

Using the funds he obtained from the Extrasensory and Supernatural Research Council, he employed a randomly selected group of witch doctors and took them to the desert island of Dira in the Indian Ocean. The witch doctors were given the instruction: Make it rain as often and as heavily as possible. Sprout observed the witch doctors over a period of some months. He measured the energy they put into rain-making by the number of calories consumed in dancing per hour. Rainfall was monitored during the 24-hour period which followed the rain dance.

Before you look up Sprout's paper try to imagine what outcomes the experiment might have produced. Three potential outcomes are summarized in Figure 3.10 a, b and c. Consider each of them in turn and say what conclusions you might draw about Sprout's hypothesis that witch doctors can make it rain.

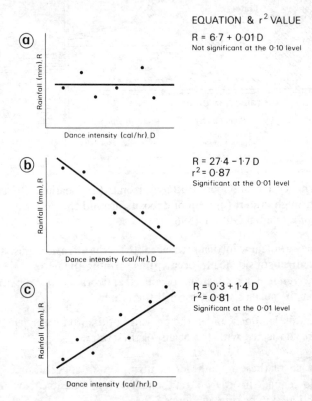

Figure 3.10 Possible outcomes of Sprout's investigations into the effect of witch doctors on rainfall
Source: after Sprout 1984

Deduction and prediction

The critical rationalist view of science is sometimes referred to as the 'deductive method' or even the 'hypothetico-deductive method'. Deductive reasoning has a role to play in the critical rationalist approach, but, as we have already seen, deductive reasoning is not exclusively confined to the Popperian account. Explanation is generally regarded as being deductive in character. Clearly we need to consider just how deductive logic is used in the critical rationalist approach to science and why it is regarded as of such importance.

Popper (1972a, p.59; 1974, pp.122–123) accepts that the covering law model describes the way in which explanations are constructed in science. However, he extends the pattern of deductive logic to other areas of scientific work. A key element in his account is the idea that theories and hypotheses should be *tested*. How are such tests to be constructed? What do such tests entail? As we shall see, testing involves the scientist in making *predictions*. Deductive logic has an important role to play in this process.

To test a theory requires an examination as to whether it can be used to make some prediction about the world, and whether this prediction appears to correspond to the way things actually are. For the test to be an informative one, the prediction must follow as an inevitable and inescapable consequence of the theory. Only then, if the prediction fails, can we say that the theory is false. How can predictions be made so that they have this property of following inescapably if the theory is true? It can be done using the same kind of logic already illustrated by the covering law model.

The covering law model consists of the asociation of two kinds of statement: a universal law obtained from some theory, and a statement about particulars (initial conditions), which together allow the theory to be applied to the specific event. The explanation is simply a statement describing the event which follows from these two premises. If the scientist is interested in explanation, then the deductive logic provided by the covering law model makes the explanation an inevitable consequence of holding some theory (which is presumed to be true). If, rather than being concerned with explanation, the scientist is interested in testing the truth of the theory itself, then deductive logic is applied in exactly the same way. Instead of regarding the conclusion of the deductive argument as an explanation it can be viewed as a statement of what must follow if a theory is true and a particular set of initial conditions apply. Because the conclusion follows logically, the prediction can be used as a test of the theory since the theory cannot be true if the prediction fails to correspond to observations. Deductive logic provides the link between the trial solution and the stage of error elimination shown in Popper's schema of critical rationalism (see above). These ideas may be explained more fully by looking at the material of the case studies.

Case Study 3.1 illustrated how a prediction is used to test a theory. Simberloff argued as follows:

1. If island area has a direct effect on species number, and
2. if the area of a mangrove island consisting of a single habitat type is reduced, then

3. species number should decline when mangrove island area is reduced.

Compare this pattern of reasoning with that of the covering law model described in Chapter 1. By using deductive reasoning it is clear that we can make some judgement about the law implied by Statement 1, and thus the theory on which it is based, by looking to see whether the prediction (3) is upheld. The results presented in Case Study 3.1 seem to corroborate the idea that area can have a direct effect on species number.

Verification and falsification

An important assumption of critical rationalism is that there is no sense in which hypotheses can ever be verified. It is argued that hypotheses can only be refuted. This point may seem rather difficult to accept because many people think science is in the business of providing proof for its theories. Unfortunately, proof of a theory is never possible. The basis of the assertion about verification and falsification can be explained by reference to Case Study 3.4. It illustrates a very important logical point.

Consider what it would mean for Sprout's hypothesis that witch doctors make it rain if Outcome a (Figure 3.10a) was observed. If witch doctors dancing make it rain we would deduce that the more energetically they danced the more it should rain. If Outcome a was the result of the experiment, then it is clear that we would be inclined to reject Sprout's hypothesis. What would it mean if we observed Outcome b in the experiment (see Figure 3.10b)? This result would also be awkward for Sprout's hypothesis. Whatever effect witch doctors have on rainfall, they certainly do not make it rain by dancing. Once again we could reject the hypothesis. Now consider what could be concluded if the experiment produced Outcome c (Figure 3.10c). Would you accept Sprout's hypothesis as *proved to be true?*

We hope you would not accept Sprout's hypothesis as true. Moreover, we hope that your reasons for not accepting it amount to more than 'I know witch doctors don't make it rain'. There are good logical reasons for not accepting the hypothesis, even if the data of Outcome c were genuine.

The reason for not accepting that Outcome c proved the hypothesis to be true, is that, although it is consistent with the hypothesis being true, it is also consistent with many others. For example, it is consistent with the hypothesis that the positive association is merely accidental. It is also consistent with the hypothesis that the witch doctors dance more energetically when their feet are itchy. Their feet get itchy when relative humidity goes up, and relative humidity goes up just before it rains. The example demonstrates the *asymmetry* between verification and falsification. No number of positive results could ever establish Sprout's hypothesis as true, even if the experiment was repeated elsewhere with different witch doctors. In contrast, one single negative outcome would refute the hypothesis.

We have devised this example of witch doctors so that you would be predisposed to reject the hypothesis even though there might be evidence in favour of it. If we had used more respectable data you might be less inclined to act so boldly. Nevertheless, the same logical result would hold. In order to show that this is so, consider the other case studies in this chapter.

In Case Study 3.1, which dealt with the problem of island species richness, the outcome of the experiments were consistent with various predictions which could be deduced from the hypothesis that area does have a direct effect on species number. The islands recovered their equilibrium number after fumigation. When island area is reduced, species number declines. Moreover, the control island (see Figure 3.3) showed that outside factors such as climate could not have caused species numbers to fluctuate, because its numbers remained constant throughout the study. The only way in which the experimental islands differed from the control was that their area was changed. However, none of these results constitute proof of the hypothesis that island area does have a direct effect on species numbers. If we cannot think of alternative explanations for the outcome of the experiments, all we can say is that the hypothesis has withstood our efforts to refute it. The best we can say is that the evidence that we have *corroborates* (i.e. is logically consistent with) the hypothesis. The results are also consistent with other hypotheses, such as that the change in island area affected microclimate and this resulted in extinctions. Simberloff's hypothesis about the direct effect of island area is, however, scientific because it can be refuted if it is false. In the experiment described in Case Study 3.1, species number could have remained unchanged despite the reduction in island area. Merely because the prediction from the theory appears to be confirmed there is no logical reason to believe that the theory is true; nor is there any way in which its truth can ultimately be established. According to the critical rationalist, scientific knowledge has nothing certain about it.

Falsification and falsifiability

The asymmetry between verification and falsification is a simple logical proposition. It cannot be avoided. Although it must guide the way in which scientific judgements are formed, in practice it cannot be applied in a cavalier manner. As we saw in Chapter 1, science involves judgement as well as logic, and the same applies here. In order to examine the roles of logic and judgement let us consider what it means to say that a theory or hypothesis is falsified.

Suppose we had conducted an experiment to test the predictions from a theory and the results of the experiment showed the predictions were not as we expected, we might say that the experiment refuted the theory. Such a decision would assume, of course, that the results of the experiment are true. If we try to proceed in this way then we are faced with a paradox. A key feature of the critical rationalist account is the idea that there is no way in which the truth of any statement can be established. If this is so then the idea of falsification as a basis for judgement appears to be unworkable in practice. The paradox is resolved, however, once we consider the different roles of logic and judgement in the scientific process.

There is, Popper (1972a, pp.86–87; 1983 pp.xix–xxv) argues, a distinction between falsifiability and falsification. Although scientific statements must be *falsifiable* (i.e. capable of being refuted), whether or not they are *falsified* (i.e. actually refuted) by a particular test is quite another matter. Whether a theory is falsified depends upon the particular judgement we form about results of the critical experiment. All scientific work, Popper argues, is speculative. Not only are our theories never finally established, but the results of observations and experiments are also tentative. As we have seen, the process of observation and experimentation is governed and guided by theory. It depends on theoretical ideas, and involves abstract concepts, even though it apparently concerns particular events. The truth of an observation can never finally be established. We must be as critical of the results of an experiment, or set of observations, as of the theory we seek to test. However, once a judgement has been formed about the reliability of a set of observations one can proceed to examine the implications of the results for a theory.

In order to examine the roles of logic and judgement in science let us consider the kinds of decision that a scientist may face if a set of experimental results contradicts a theory. Sparks (1962) suggests that the scientist can adopt one of three strategies:

1. reject the theory
2. reject the refuting evidence
3. develop a new or modified theory which accounts for the successful predictions of the old theory, and at the same time explains how the new observations arose.

Each of these strategies is illustrated in the investigations of the age of the outermost moraine at Storbreen (see Case Study 3.2). Consider Matthews's 1750 end-moraine hypothesis. Merely because he was unable to refute the hypothesis on the basis of the lichenometric evidence, it is misleading to conclude that the moraine at Storbreen does indeed date from 1750. Griffey and Matthews (1978), for example, went on to date a moss layer (site 7) beneath the moraine and obtained a series of radiocarbon dates which suggested an age older than that required by the 1750 end-moraine hypothesis. Once faced with such results some judgement had to be made about which set of observations were the most reliable. If the age of the moss layers beneath the end-moraine is 532 or 664 years BP the moraine must be older than the mid-eighteenth century. On the other hand, if the lichenometric evidence is correct, then the radiocarbon dates are wrong. If we are unable to refute either possibility, then some other hypothesis must be found that would reconcile these apparently conflicting observations. Which strategy should be adopted?

Griffey and Matthews choose to reject radiocarbon dates, suggesting that:

A combination of the possibility of the erosion of the peat above the upper moss layer during the formation of the moraine ridge, the possible cessation of peat accumulation at the sample point after the growth of the upper moss layer, and the accuracy limitations of the ^{14}C technique, means that the maximum is possibly the mid-eighteenth century one suggested by previous studies ... (pp.85–86)

When evaluating the conclusions of Griffey and Matthews, it is interesting to note that these workers did not develop an idea which might explain both sets of observations. The outermost moraine could, for example, represent the terminal position of

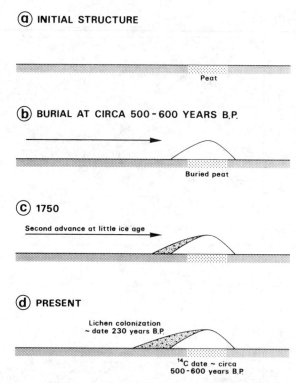

Figure 3.11 Hypothetical model showing effect of two ice advances at Storbreen
explaining anomalous lichen and radiocarbon dates

two advances (see Figure 3.11). The first advance buried the soil at about 600 years
BP. The second deposited fresh material on the proximal slope of the moraine in the
mid-eighteenth century, and it is this to which the lichenometric dates refer.

In more recent work Matthews (1980) has presented further evidence that may
cause one to doubt the radiocarbon dates presented in the earlier work with Griffey.
Matthews (1980) found that, in a buried soil, radiocarbon ages increased for
successively deeper layers within the deposit (see Figure 3.12). This is perhaps hardly
surprising since organic matter is incorporated into the soil from the surface. Once
soil development is arrested the surface horizons will always appear more recent.
Matthews (1980) argues that the ages obtained in the earlier study are misleading
because they were the average of several layers which were not themselves representative of the top-most horizon of the deposit. Further work by Matthews and Dresser
(1983) has substantiated the existence of a steep increase in age with depth.

Whether one agrees with Matthews's subsequent claims or not, his change of
strategy is clear. He looks at the new problem situation produced by the various tests
of the 1750 end-moraine hypothesis, and develops a theory which can explain just
how the lichenometric and radiocarbon evidence have come into apparent conflict.

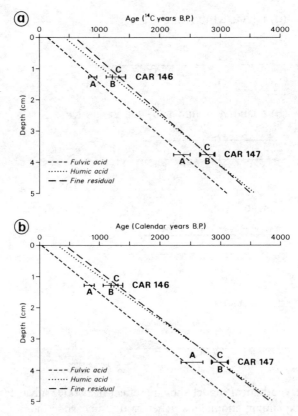

Figure 3.12 Age–depth relationships within the surface horizons of a soil: (a) ^{14}C
years plotted against mean sample depth for various organic fractions
from the soil; (b) calendar years plotted against mean depth for the
same three fractions
Source: after Matthews 1980

The new theory about the age–depth profile of soils resolves the problem and the
1750 hypothesis is once again secured.

Multiple working hypotheses
In the critical rationalist model the idea of hypothesis testing is extremely important.
Yet the account given so far has simplified matters because it implies that the scientist is engaged in testing a single hypothesis. This is rarely the case. Usually the scientist is in the situation of deciding *between competing* hypotheses. The stage of error
elimination consists of making just those observations, or performing just those
experiments, that help to decide between the competing theories.

If we consider Case Studies 3.1 and 3.2, then it is clear that in each study two hypotheses are being tested against each other. In the case of island species number, the hypothesis that is stated explicitly is that area has a direct effect on species numbers. The unstated, but nevertheless competing hypothesis is that area has no direct effect. This second hypothesis may not seem very exciting. It seems to be no more than a 'catch all'. It serves an important function, however. In order to appreciate its importance we must recall just what it is that is supposed to make a theory 'scientific'.

According to the critical rationalist, a statement or a theory is considered scientific if it is testable (i.e. capable of being falsified). In order to have this characteristic the statement must *exclude something*. That is, it must allow the possibility of making observations that would conflict with the theory. If such observations are made, then the theory might be refuted and we might move on to consider the alternative hypothesis.

In Case Study 3.1, for example, the test of the theory that island area has a direct effect on species number involves trying to set up an experimental situation that would allow the theory to fail if it were false. Thus all other factors which might affect island species richness are held constant. The influence of environmental heterogeneity is removed by using islands consisting of a single habitat type (only mangroves). Simberloff ensures that changes in species number on the experimental islands cannot be attributed to seasonal effects by studying an island whose area was not changed but which nevertheless maintained a constant species number. By using such *experimental controls*, changes in species number on the experimental islands can be attributed only to a change in area. Although, as we have seen, the experiment cannot be used to confirm the theory about the direct effect of area on species number, if the theory is correct then area cannot change without affecting species number. If in the experiment species number was not found to change, then because of the way in which the experiment was set up the theory about the direct effect of area must be considered false. Thus the theory is tested by trying to observe something which it says cannot happen.

In the case study dealing with island species number, the competing hypothesis was very general in character. Case Study 3.2 illustrates how an initially rather vague alternative hypothesis becomes more explicit as work develops. In the initial study Matthews (1977) tested the hypothesis that the end-moraine at Storbreen dated from 1750 against a general hypothesis, that it did not. Further experimental testing suggested that the moraine might be older. The problem situation then consisted of deciding between a more clearly stated set of opposing views.

The way in which the scientist uses the results of experiments and observations to make judgements about the merits of alternative theories is perhaps most easily seen in Case Study 3.3 concerning the origin of the crater in Arizona. Indeed, Gilbert (1896) used the problem of the origin of Coon Butte to illustrate the so-called *method of multiple working hypotheses* first proposed by Chamberlain (1890).

Gilbert's paper is instructive because it illustrates the role that field-work should play in the environmental sciences. He does not proceed by collecting 'data' as in the classical tradition. Instead, he attempts to make just those observations which help to decide between the competing hypotheses. Thus, if the crater resulted from meteorite impact, then the presence of an iron mass below the surface should be detectable by looking at deflections of a magnetic needle. Alternatively, if it resulted from an explosion, then the volume of material outside the crater should equal the void which the explosion is supposed to have created, and so on. The character of the field observations are not only dependent on theory (effect of iron on magnetic needles etc.) but are also guided by the character of the theories which are being tested.

In our introduction to this chapter the critical rationalist approach was summarized by the following simple schema:

$$P_1 \rightarrow TT \rightarrow EE \rightarrow P_2$$

This simple model can be rewritten (cf. Popper 1972a, p.243) in order to represent the situation that is more usual in scientific work, where several hypotheses or tentative theories are put forward:

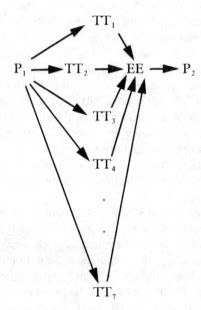

The aim of science is to solve problems. Each of the tentative hypotheses represents a feasible solution, and a critical investigation may enable a judgement to be made between them. The parallels between this account of science and the method of multiple working hypotheses is striking. Indeed, the similarity has been acknowledged by Popper himself (see Schilpp 1974, p.1009 and p.1187 note 80). Popper's account of science adds a new and important dimension to the writings of Chamberlin and

Gilbert, and appears to establish their ideas as a key element in the methods employed in the natural sciences (cf. Baker and Payne 1978). An excellent example of the method in operation is provided by the work of Battarbee et al. (1985) on the causes of lake acidification in Galway, Scotland. They examine the hypothesis that it is due to acid precipitation against the competing hypotheses that it is due to heathland regeneration, afforestation or long-term post-glacial natural acidification.

Falsification and ad hoc theories

Popper argues that it is through criticism that false ideas can be eliminated, and theories gradually improved. However, although refutation may cause the scientist to modify and refine his theories, not all changes to a theory are equally acceptable. In particular, Popper (1972a) argues that when a scientist faces a potential refutation of his theory he should avoid making *ad hoc* changes simply as a means of avoiding falsification.

The most important thing about a scientific theory is that it is testable. Theories can be modified in any way, providing this characteristic of potential falsifiability is retained or even improved. An ad hoc modification to a theory is any change which is introduced merely to protect the theory from refutation and does not improve its testability. The importance of avoiding ad hoc changes can be illustrated in the following example.

Suppose a scientist developed a theory which led him to propose the law that:

water boils at 100°C

If he conducted an experiment which involved boiling water at high altitude he would find that the law was refuted. One strategy which the scientist could adopt in order to save the law, and the theory on which it is based, would be to add the clause:

... but not at high altitudes.

This modification to the theory would be regarded as an ad hoc change because, although it prevents refutation, it does not improve the ability of the theory to be tested. All refutations could be treated in this way until the list of exception clauses becomes so long that the theory is unmanageable.

Faced with the refutation of the theory about boiling water the scientist could adopt another approach. He could ask what it is about high altitude that causes water to boil at less than 100°C. A new theory may then be developed, this time involving ideas about the way in which air pressure affects boiling point. This modified theory becomes more testable than the old one, for now it includes reference to a wider range of situations which may potentially refute it. Although the theory has been modified, its testability or *empirical content* has increased. Such changes are not regarded as ad hoc and are to be encouraged.

Although the drawbacks associated with making ad hoc changes to theories are obvious, the scientific literature contains many examples of this strategy being employed to save a theory. Consider, for example, the response of Griffey and Matthews (1978) to the observation that the radiocarbon dates obtained below the end-moraine at Storbreen were older than the supposed date of 1750 (see Case Study 3.2)

Griffey and Matthews (1978, p.85) argue that the dates obtained at site 7 (532 ± 40 years, and 664 ± 45 years BP, see Table 3.1) demonstrate a Little Ice Age maximum. This is despite the fact that the dates should be nearer to 230 years BP. They attribute the disparity in the radiocarbon dates to erosion of the peat above the dated moss layer, the possible cessation of accumulation of peat after the growth of the moss layer, and the limitations of the accuracy of the radiocarbon technique. At the other dated site (site 6), they argue as follows:

> The interpretation of the ^{14}C date ... is not so straightforward. Three interpret-
> ations can be considered: the moraine ridge was formed during the expansion
> episode dated at Lierbreen [circa 1350 years BP]; or a somewhat more recent
> expansion episode; or the 'Little Ice Age'. (p.86)

Of the three possibilities they accept the Little Ice Age hypothesis. Once again their choice is influenced by the possibility of some erosion of the profile above the sample point, and the limitations of the ^{14}C technique.

In evaluating Griffey and Matthews's (1978) work it becomes clear that the 1750 end-moraine hypothesis is retained for reasons which are not altogether convincing. The rejection of the radiocarbon dates at Storbreen seems arbitrary because they are accepted at the other forelands investigated. Indeed, on the basis of these sites there would seem to be fairly general evidence that the 1750 end-moraine hypothesis should be refuted, if their radiocarbon dates are accepted.

Although the reasons given by Griffey and Matthews (1978) for retaining the 1750 end-moraine hypothesis at Storbreen could be regarded as ad hoc, the argument which Matthews later developed to support this conclusion is much more acceptable. Matthews (1980) argued that soils show an age–depth relationship. The deeper the radiocarbon sample the older its apparent age. Since the samples used in the previous study (Griffey and Matthews 1978) were obtained from relatively thick layers, their ages are misleading. *All* earlier dates are now considered as suspect, and all are now regarded as consistent with burial in the Little Ice Age. Thus, in the later work, it is clear that the 1750 end-moraine hypothesis is no longer retained for ad hoc reasons. Its apparent conflict with the radiocarbon data is resolved by the development of a new, testable idea about the age–depth profile of soils.

Corroboration and verisimilitude

As we have seen, the concept of proof, or verification, has no place in critical rationalism. If an experiment or set of observations appears to correspond to the predictions of a theory, then all that can be said is that the evidence *corroborates* the theory. Popper chooses the word corroboration deliberately. He wishes to convey that, although the evidence is consistent with the theory, this cannot be taken as proof that the theory is true. A theory is corroborated so long as it stands up to critical tests. If proof is never possible, what role does the idea of truth play in science?

Although the effort to establish true theories regulates the operation of science, there is no way the scientist can ever be certain that a true theory has been developed. Despite this limitation, Popper argues that progress in science is possible. Using the method of conjecture and refutation, hypotheses can be improved and false ideas eliminated. Thus there is a sense in which science can progress. Progress is marked by the development of more precise and comprehensive accounts of reality. Although there is way that the truth of a theory can finally be assured, theories can be made better approximations to the truth by the process of trial and error elimination. The aim of science, Popper suggests, is to produce theories of greater *verisimilitude* or higher truth content.

The idea of verisimilitude arises directly from Popper's acceptance of the correspondence theory (see Popper 1972b, pp.47–60). According to the correspondence theory a statement is considered true or false according to whether it does or does not correspond to the facts. Verisimilitude is a measure of the degree to which a statement corresponds to the truth. The idea is very simple to grasp if it can be accepted that, although two statements may be false, one could represent a better approximation to the truth than the other. Popper (1972b, p.103; 1983, p.57) considers that an indication of the verisimilitude of a theory may be gained by an assessment of the extent to which it has been corroborated. He cautions, however, that corroboration should not be taken as a measure of verisimilitude:

> The degree of corroboration is a guide to the preference between two theories at a certain stage of discussion with respect to their then apparent approximations to the truth. But it only tells us that one of the theories offered *seems – in the light of discussion – the one nearer the truth.* (Popper 1972b, p.103)

Popper suggests that the situation which the scientist faces is described by the following extract from Xenophanes:

> The gods did not reveal, from the beginning,
> All things to us; but in the course of time,
> Through seeking we may learn, and know things better.

> But as for certain truth, no man knows it,
> Nor will he know it; neither of the gods,
> Nor yet of all the things of which I speak,
> And even if by chance he were to utter
> The final truth, he would himself not know it;
> For all is but a woven web of guesses.
> (quoted by Popper 1974, p.26)

All may indeed be a woven web of guesses, but there is, nevertheless, a rational basis for the judgement that some guesses more accurately represent the world than others. This rational basis is provided by the critical investigation of theories.

Critical rationalism and physical geography

The approach to science represented by critical rationalism has been acknowledged by a number of geographers, including: Bird (1975), Moss (1977), Wilson (1972), and Marshall (1982) (see also Platt 1964). In physical geography the 'deductive method' has been represented as one on which increasing emphasis has been placed. Anderson and Burt (1981), for example, describe the move away from historical explanation represented by such schemes as Davis's cycle of erosion, and conclude:

> The trend towards a more detailed consideration of process-response systems was therefore accompanied by the abandonment of both the temporal mode of explanation and the inductive approach with which it was associated. Geomorphology thus moved away from being essentially a classificatory procedure and adopted a more truly scientific attitude based upon the use of the deductive method. (p.4)

However, despite such wide acknowledgement, many of these comments on the deductive method completely obscure its main ideas.

Thus, Anderson and Burt (1981) summarize the deductive method by means of a diagram derived from Harvey (1969), see Figure 3.13. If this diagram is considered in the light of Popper's views and Figure 3.1, then it is clear that it completely misrepresents the deductive model. Verification, for example, has no role to play in the deductive route to scientific explanation. Laws and theories, according to the deductive model, are not obtained by 'successful' verifications.

The misrepresentation of the deductive approach goes further. Anderson and Burt go on to refer to Chorley's (1966) summary of the main branches of geomorphology (see Figure 3.14), and suggest:

> Chorley's scheme correctly isolated theoretical work from the data collecting stage, this division being necessary if the deductive approach is to be followed correctly. (1981, p.8)

As we have seen a major assumption of the deductive model is that theory and observation *cannot* be isolated from each theory. All observation presupposes some theory. To suggest that data accumulation, whether it be in the field, laboratory or armchair, is an acceptable end in itself, is entirely at odds with the 'more truly scientific attitude' of the deductive approach.

The confused account of Anderson and Burt (1981) reflects a widely held belief that the inductive and deductive methods represent alternative, and equally valid,

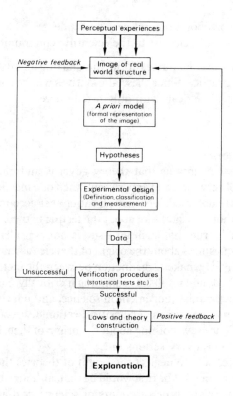

Figure 3.13 A misrepresentation of the deductive method
        Source: after Harvey 1969

routes to explanation. If physical geography is beginning to use the deductive method, then the implication is that inductive strategies must be abandoned. Inductive and deductive approaches do not represent alternative but *conflicting* accounts of scientific method.

If physical geographers are adopting the deductive approach in their work, then a more complete understanding of the implications of the ideas which lie behind it is required. This chapter has introduced much of the background to the approach. In order to establish a complete picture we must now turn to what some people see as its limitations.

The limitations of critical rationalism

Despite its wide acceptance, the critical rationalist view has itself been criticized. We will summarize these criticisms by reference to three problem-areas:

– the origin of theories

– the concepts of corroboration and verisimilitude
– the relationship of the scientist to the scientific community

The second and third problem-areas form the starting points for quite different accounts of scientific practice. Since these alternatives will be considered in the next three chapters, they will be dealt with only briefly here.

1. The origin of theories

The critical rationalist maintains that theory governs and guides observation. If theory does precede observation and experiment, then one may legitimately ask how theories themselves are obtained. The critical rationalist account of science is often criticized because, although it says a lot about the testing of theories, it says very little about their origin. It is true that such an issue is not considered. In fact, Popper (1972a, p.31) rejects questions about the origin of theories as matters for psychology rather than philosophy. He makes a distinction between the methods of conceiving a new idea and the methods and results of examining it critically. For him, methodology is about the logic which guides judgement in science, and not the subjective factors which lead to creation of new ideas. The critical rationalist would therefore accept that his account of science says nothing about the origin of theories, but would argue that the deficiency is not a very serious one.

In defence of the decision to neglect the origin of theories the critical rationalist might make three main points. The first would be that, although the origin of theories is not discussed, none of the other accounts of science deal adequately with this problem either. The classical model apparently shows that theory is the outcome of 'unbiased' observation, but the rationalist would argue that such an account is mistaken. Theory cannot be divorced from observation. Even if inductive logic exists, theory controls observations from the outset. Secondly, it would be argued that, in any case, an account of the way in which ideas are conceived is irrelevant to the judgement which is eventually made about their correspondence to reality. The significance attached to an idea does not depend on its origin, but on its ability to stand up to serious criticism. If pressed on the matter of the origin of theories the critical rationalist would argue that theories can be generated in almost any conceivable way. Some people (see, for example, Gilbert 1896 and Chapter 8) may choose to argue by *analogy*. Others may rely upon flashes of inspiration or hunches. Some may even attempt to salvage something from the ideas of inductive reasoning, by regarding it as a method of 'hypothesis generation'. None of this, the critical rationalist would argue, is science. The third point which the critical rationalist would make is that the problem of the origin of theories is irrelevant. Only when ideas are subjected to critical investigation does science really begin.

2. The concepts of corroboration and verisimilitude

Although the truth of theories cannot be established absolutely, the critical

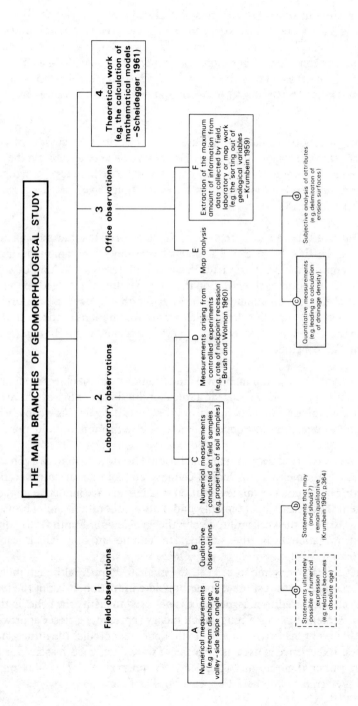

Figure 3.14 The main branches of geomorphology
Source: after Chorley 1966

rationalist maintains that by the process of trial and error theories can be improved and achieve a closer approximation to the truth (i.e. greater verisimilitude). Some have suggested that such an idea is difficult to support. For example, Newton-Smith (1981, chapter 5) considers what rational basis there is for making judgements about the verisimilitude achieved by a theory. According to Popper, Newton-Smith notes, the scientist is guided by the degree of corroboration of a theory, where corroboration is

> . . . a concise report evaluating the state . . . of the critical discussion of a theory, with respect to the way it solves its problems; its degree of testability; the severity of tests it has undergone; and the way it has stood up to those tests. Corroboration (or degree of corroboration) is thus an evaluation *report of past performance* (Popper 1972b, p.18)

Newton-Smith (1981) considers this idea unsatisfactory.He attacks Popper's thesis on two grounds: first because the degree of corroboration cannot be used as a measure of approximation to the truth, and second because the kind of argument used by Popper amounts to the assertion that theories are to be judged on the basis of experience. The argument amounts to the assertion that if they have performed well or better in past tests then they must be closer approximations to the truth than any rival. Such an argument, Newton-Smith (1981) reasons, is surely *inductive*. As a consequence he concludes:

> . . . if we admit these grand high-level inductions we cannot object to all inductive argumentation *per se*. If we concede a role to induction here there is no reason not to admit inductive arguments right from the start. If we do this we lose what was unique and interesting in Popper: namely, the jettisoning of induction. (p.70)

O'Hear (1980, p.67) reaches a similar conclusion. He suggests that, as a consequence of the kind of argument used, Popper cannot complain about corroboration and theory testing being taken by his less cautious readers to have inductive implications. To the extent that truth is unobtainable, and that corroboration cannot be used as an effective measure of the verisimilitude of a theory, Newton-Smith (1981) questions whether it is rational for the scientist to pursue an aim which cannot be realized.

Such criticisms have some force. Yet, to be fair to Popper, although judgements about theories appear to rest upon something like induction, it is not the inductive logic of the classical tradition. Experience is not used to justify the belief in the truth of a theory. Statements concerning which theory represents a closer approximation to the truth are not made with any belief that one may predict the future success of the theories. Experience is used as a means of providing 'good reasons' for holding one theory rather than another (Popper 1893, pp.xxxv and 71). Why might one theory be preferred to another? Popper argues as follows:

When faced with the *need to act* on one theory or another, the rational choice was to act on that theory – if there was one – which had so far stood up to criticism better than its competitors had: there is no better idea of rationality than that of a readiness to accept criticism; that is, criticism which discusses the merits of competing theories from the point of view of the regulative idea of truth. Accordingly, the degree of corroboration of a theory is a rational guide to practice. Although we cannot justify a theory – that is justify our belief in its truth – we can sometimes justify our *preference* for one theory over another ... (Popper 1976, pp.103–104).

Thus to appeal to experience as the basis for judgement is not to admit that induction is either necessary or possible. Although truth of theories cannot finally be established, rational decisions can be made about their relative merits.

3. The scientist and the scientific community

Although the critical rationalist stresses the role that logic plays in science, acceptance or rejection of a theory is not a matter of logic alone. As we have seen, there is a distinction between falsifiability and falsification. All scientific statements must be falsifiable, but whether they are falsified in a particular instance is quite another matter. Theories are judged according to the view which the scientist and the scientific community form of the basic statements produced by their programmes of experimentation and observation. To some, notably Kuhn and Feyerabend, the existence of such elements is sufficient to undermine the claim of any kind of rationality in science at all. For them, progress in science is not to be understood in terms of a method grounded in reason. Rather, it is to be explained in terms of the sociological processes within the scientific community.

Kuhn and Feyerabend reinforce their attack on critical rationalism by pointing out that it does not fit the history of science itself. As a result they try to develop an account of science which is firmly based on scientific practice rather than on some elitist ideal which is never realized. While Popper seeks to prescribe what scientists *should* do, Kuhn, Feyerabend and others have sought to describe what scientists *actually* do.

The divisions between the views of science represented by Popper, Kuhn and Feyerabend are deep. An assessment of their relative merits cannot, however, be attempted until these opposing ideas about science are explored more fully. This task will be undertaken in the chapters which follow.

Critical rationalism: interim review

As a response to the problems facing the classical model, Popper has gone some way to providing what Russell (1961, and Chapter 2) demanded, namely an answer to

Hume '. . . within a framework of philosophy that is wholly or mainly empirical'. He has done so by showing that, despite the tentative nature of all knowledge, ideas about reality can be sifted and assessed rationally. Although beliefs in the truth of theories can never be justified, there is a framework in which rational judgements can be made. This framework is provided by the method of trial and error, or conjecture and refutation. Some have sought to attack this view, and we will examine their arguments elsewhere. At this stage, however, it is important to review the image of science provided by Popper and to compare it with the classical view.

In contrast to the secure image of scientific knowledge painted by the classical tradition, Popper describes his view of knowledge as follows:

The empirical basis of objective science has thus nothing 'absolute' about it. Science does not rest upon solid bedrock. The bold structure of its scientific theories rises, as it were, above a swamp. It is a building erected on piles. The piles are driven down from above into the swamp, but not down to any natural or 'given' base; and if we stop driving the piles deeper, it is not because we have reached firm ground. We simply stop when we are satisfied that the piles are firm enough to carry the structure, at least for the time being. (1972a, p.111)

Whether the Popperian structure is secure enough to withstand the attacks of Kuhn and Feyerabend remains to be seen.

CHAPTER 4:
THE KUHNIAN VIEW

'Man is the measure of all things'

Introduction

When faced with two competing theories some people believe the scientist can usually provide good reasons for choosing one rather than the other. By means of logic, experiment and observation, the scientist makes some judgement about the match or correspondence of the theory to reality. As we have seen, people who hold such views are described as rationalists. Their major claim is that judgement in science has a rational basis. Despite the profound differences between them, the classical and critical rationalist views of science share the common assumptions of rationalism. By no means all of those who have considered the nature and methods of science agree with such views, however. Kuhn and Feyerabend, amongst others, have attacked the idea that science is an entirely rational activity. They claim that there are other forces, besides reason, which shape scientific judgement. For these writers the categories 'true' or 'false' do not mean anything as absolute as the rationalists maintain. Rather, what is accepted or rejected depends mainly on the social or intellectual outlook of the scientist or scientific community. The ideas of Kuhn and Feyerabend represent a form of *relativism* (see Chapter 1). This is because they believe that judgements are made relative to some accepted set of community standards or norms, rather than wholly by reference to an external reality.

In order to introduce the ideas of relativism let us consider why Kuhn and Feyerabend have rejected reason as the basis of judgement. A clue can be found in Kuhn's reaction to Popper's ideas about falsification:

> . . . no theory ever solves all the puzzles with which it is confronted at a given time; nor are the solutions already achieved often perfect . . . If any and every failure to fit were grounds for theory rejection, all theories ought to be rejected at all times. On the other hand, if only severe failure to fit justifies theory rejection, then the Popperians will require some criterion of 'improbability' or of 'degree of falsification'. (1970a, pp.146–147)

Kuhn doubts whether a criterion of 'improbability' or 'degree of falsification' can be found. Yet, what stops the scientist rejecting a theory every time an anomaly appears? Kuhn argues that judgements are formed about theories on grounds other than reason based on experience. Sociological and psychological factors are the major consideration. Dury (1978, pp.270–271), in his analysis of the medium-term trends in geomorphology, has illustrated how such forces might act in funding and publishing. Chalmers (1982) summarizes Kuhn's view very well in terms of Protagoras's dictum: 'man is the measure of all things'.

Both Kuhn and Feyerabend appeal to history in order to show that forces other than reason have shaped scientific development. In this and the chapters which follow we will examine the ideas of relativism and some ways in which people have reacted to them.

Kuhn: consensus and schism in science

A study of the history or structure of most scientific disciplines would probably reveal periods of both consensus and schism in the scientific community. At one time or place scientists appear to share a common world view and have similar goals and values. In other situations there is division. Images of the world compete and conflict. There is little common ground until one world view can achieve dominance, so that consensus is restored. A study of the history of many sciences might suggest that these patterns are the essence of scientific progress.

For a rationalist like Popper such patterns of change would be viewed as the result of the process of conjecture and refutation. Images of the world change as their flaws are discovered. Kuhn, in his *Structure of Scientific Revolutions* (1962, 1970a; see also Barnes 1982), has also sought to explain the dynamic character of science. He argues that knowledge is judged by reference not to its correspondence with reality, but to some framework of received opinion.

Kuhn (1970a, p.10) suggests that much of science can be characterized as representing some kind of consensus. At such times research is firmly based on past scientific achievements (e.g. some classic work, or theory). These achievements are taken to define the legitimate problems and methods of the research field. Moreover, such achievements are sufficiently open-ended to provide scope for further work. Using the accepted stock of concepts and techniques, past successes may be continuously refined. Kuhn describes such achievements, which simultaneously gather allegiance and provide a framework for further work, as *paradigms*. In choosing such a term he wishes to suggest:

> . . . that some accepted examples of scientific practice – examples which include law, theory, application and instrumentation together – provide models from which spring particular coherent traditions of scientific research. (1970a, p.10)

Kuhn describes those periods when research proceeds on the basis of some coherent tradition as *normal science*.

Work during periods of normal science is characterized by something like the following sequence of steps:

1. the learning, understanding and acceptance of some dominant paradigm by the individual scientist;
2. the recognition of puzzles posed by the paradigm;
3. the attempt to solve those puzzles using the concepts and techniques provided by the paradigm;
4. the acceptance or rejection of the proposed solution by the community of scientists.

Kuhn (1970a) explains that the history of science shows that scientists are not normally engaged in trying to overthrow theories and to develop new ones. Instead scientists tend to accept established ideas and use them to solve puzzles. Kuhn uses the term puzzle deliberately. He wishes to convey the idea that during periods of normal science the scientist is not engaged in solving deep or fundamental problems which concern the validity of the paradigm itself. Kuhn regards the problems tackled by normal science as essentially routine, because it is expected that they can be solved by merely applying what the scientist has already learned and accepted. Under the ruling paradigm the existence of a solution to such puzzles is generally taken for granted and the success of the normal scientist is judged according to whether he has successfully applied the paradigm to his problem. Failure generally reflects on the competence of the scientist rather than on the adequacy of the paradigm itself.

Although the bulk of scientific development occurs during times of 'normal science' there are also periods of upheaval and change. Kuhn (1970a) describes such periods as *scientific revolutions*.

Scientific revolutions develop as the result of the accumulation of anomalies associated with the ruling paradigm. These anomalies are generally puzzles and issues which cannot be resolved using traditional concepts and techniques. During periods of normal science such problems are laid at the door of the scientist. During periods of revolution, however, some people are prepared to doubt the paradigm itself. An alternative paradigm may be accepted if it appears to solve the problems which brought the old one to crisis, or it may be considered that further achievements can only be made using some other world view. For a time the two paradigms may compete for the allegiance of scientists. If the new paradigm is sufficiently attractive and open-ended, then it may eventually replace the old consensus view. What was once a revolutionary movement becomes the new, institutionalized research tradition.

To emphasize the kind of transformation which a scientific revolution represents, Kuhn describes the paradigm switch by analogy with a *gestalt diagram*. A gestalt diagram is a visual illusion in which an object is at one time seen as one thing, and at other times seen as something else. Both interpretations of the diagram cannot be seen simultaneously. However, having seen both, the observer can easily switch between them. Examples of gestalt diagrams are shown in Figure 4.1. Kuhn uses the analogy to emphasize that a scientific revolution represents a radical and fundamental change of perspective. One framework for interpreting reality is totally replaced by another and the alternative views are incompatible.

Figure 4.1 Two gestalt diagrams. Each image can be interpreted in at least two mutually exclusive ways.

The Kuhnian model in practice

As a means of describing developments in science, the Kuhnian model, with its concepts of paradigms, normal science and periodic revolutions, has been widely accepted in geography (Haggett and Chorley 1967; Johnston 1979). Before the implications of these ideas are examined in detail, their character will be illustrated by the following case studies. You might also try Exercises 4.1 and 4.2.

Case Study 4.1: Plate tectonics

Jones (1974) has investigated whether Kuhn's views fit the actual changes which occurred in the earth sciences in the post-war period with the development of the theory of plate tectonics. The theory of plate tectonics grew out of what he describes as the 'mobilist paradigm', a view which contrasted with the older 'stabilist' ideas.

Jones characterizes the theory of plate tectonics as follows:

> . . . the theory states that the entire surface of the Earth consists of six major rigid plates in motion relative to each other. The plates are created along mid-oceanic ridges and are consumed beneath deep ocean trenches. At other places the plates slide past each other at what are known as transform faults. The continents consist of less dense material embedded in the upper surface of the plates on the backs of which they are trundled around the surface of the globe. (p.536)

The theory is illustrated in Figure 4.2. Jones explains that the movement of the continents described by the older theory of continental drift, which suggested that only the continental blocks are mobile, is simply a consequence of the processes envisaged by the wider, more recent theory of plate tectonics.

Jones contrasts the mobilist paradigm with earlier stabilist views by suggesting

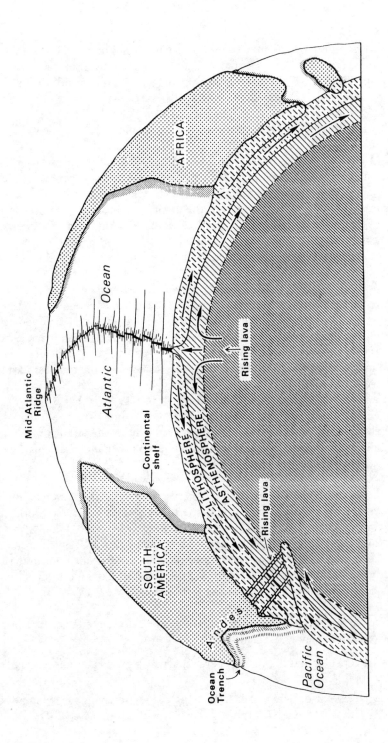

Figure 4.2 Summary of the main concepts of the theory of plate tectonics

that the essential characteristic of the former is its emphasis on horizontal movements in the earth's surface. The older views assumed that vertical movements were
the major influence on the development of the surface features of the earth.

As Hallam (1973) notes, in the mid-1960s, earth scientists switched from the
stabilist view to an approach based on mobilism (see also Wyllie 1976). Jones argues
that 'this massive swing between two mutually contradictory views is *entirely
Kuhnian*' (p.530). To emphasize his point he argues that the kind of transformation
of world view which this switch represented has many of the features of a gestalt
switch. In order to illustrate this transformation he considers the problem of accounting for animal and plant distributions on continents whose positions were once
assumed to be fixed. Working within the stabilist paradigm, scientists sought to
explain patterns by supposing land-bridges which had long since disappeared, but
with the mobilist paradigm such problems no longer arise. Instead the new paradigm
poses puzzles consisting of interpreting distributions in terms of the history of a
mobile earth.

The switch to the mobilist paradigm poses something of a problem for those
interested in the history of the earth sciences. Although it was not until the 1960s that
scientists began to accept mobilist views, the idea of continental drift had been put
forward fifty years earlier by Alfred Wegener (1912; see also Wegener 1966), and
Hallam (1973) notes that the idea had been suggested several times before. Why did it
take so long for the revolution to occur? Why, in fact, was the mobilist idea largely
ignored until the 1960s?

The answer to these questions suggested by Jones (1974) is that Wegener's theory
was premature. For a revolution to occur, not only must the new theory appear
reasonable, but there must also be some crisis with the old paradigm. What, then,
eventually precipitated the collapse of the stabilist view? In the case of plate tectonics
Jones suggests that it was *sponsorship*:

> Plate tectonics emerged in the 1960's in large part as a direct consequence of
> oceanographic research carried out in the 20 years after the Second World War. At
> the time by far the largest share of sponsorship came from the United States Office
> of Naval Research (ONR). Without this backing by the ONR much of the crucial
> sea-bed research would have been done much later, if at all. Thus in the 1960's,
> partly as a result of the external force of sponsorship, a unique group of scientists
> was confronted by a unique data set. (p.538)

Examine the idea that the acceptance of the theory of plate tectonics is a case in
favour of Kuhn's model of science.

Do you consider that the analogy between a paradigm change and a gestalt switch
is acceptable in this case?

To what extent do you consider that the paradigm change was mainly due to sociological processes operating in the scientific community in the 1960s? Alternatively, try to identify what role logic appeared to play in the process of change.

Case Study 4.2: Quantification and the study of processes in geomorphology

The character and focus of geomorphology is often presented as having changed over the last forty years or so (Anderson and Burt 1981; Pitty 1982). Once geomorphologists apparently concerned themselves largely with the interpretation of the history of landforms. Today there is greater emphasis on the study of the processes which shape the landscape. Once the analysis was largely qualitative. Today much of geomorphology is quantitative in character. Often such changes are described in terms of the Kuhnian model. Thus one often reads of events such as the 'process revolution' and the 'quantitative revolution'. Let us consider what this change in emphasis represents. We can do so by means of comparing the views of geomorphologists at opposite sides of the divide.

The qualitative, historical approach to landscape was largely based on the ideas developed by the American geomorphologist W.M. Davis. His cycle of erosion provided the conceptual framework for much of the geomorphological work in Britain and North America during the first half of the twentieth century.

Davis argued that the character of landforms was dependent on three factors: *structure*, *process*, and *stage*. In a given climate, and in relation to a base level provided by the prevailing sea level, landscapes would develop through a sequence of stages described as 'youth', 'maturity' and 'old age', the latter represented by a major erosion surface or peneplain (see Figure 4.3). Thornes and Brunsden (1977) have described how this model was used to reconstruct the history of the landscape:

> The method was simple. First, if surface or partial surfaces could be formed then they could be recognised, mapped and placed in sequential order. Second, if an area had passed through several former cycles of erosion it would contain relics of dissected erosion surfaces which, by uplift, had formed the initial surface of the new cycle. (p.120)

The method they describe was known as *denudation chronology*. Its aim was to explain the patterns in the present landscape by reference to its history, interpreted in terms of the cycle of erosion.

In 1950 the Association of American Geographers held a symposium to honour the centenary of the birth of Davis. In the proceedings of the meeting Strahler writes:

> Davis' treatment ... was completely qualitative. I do not recall having seen a

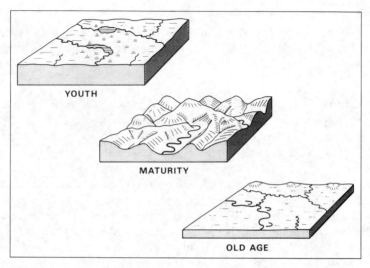

Figure 4.3 Davis's cycle of erosion

measurement of slope angle or a precisely measured slope profile in any of his publications. Neither is there a penetrating analysis of erosional processes based on mechanics of fluids or plastic materials . . . (p.209)

He adds:

Davis' treatment appealed to . . . persons who had little training in basic physical sciences . . . as part of the basis for the understanding of human geography it is entirely adequate. As a branch of the physical sciences it is superficial and inadequate. (ibid.)

As Wooldridge (1958, p.30) notes, Strahler went on to publish a series of papers whose aim was to place geomorphology on 'sound foundations for quantitative research into fundamental principles, based on fluid mechanics and fluid dynamics'. For a modern statement of the process paradigm one may consult Derbyshire et al. (1979, especially chapter 6), Brunsden and Thornes (1979), and Pitty (1982). The swing to the quantitative study of geomorphological processes was not, however, universally accepted. Wooldridge and Morgan (1959), for example, reacted to the trend as follows:

There has been a recent attempt in certain quarters to devise a 'new' quasi-mathematical geomorphology. At its worst this is hardly more than a ponderous sort of cant. The process and results of rock sculpture are not usefully amenable to treatment by mathematics at higher certificate level. If any 'best' is to result from the movement, we have yet to see it; it will be time enough to incorporate it in the subject when it has discovered or expressed something which cannot be expressed in plain English. For ourselves, we continue to regard W.M. Davis as the founder of our craft and regret the murmurings of dispraise heard occasionally from his native land. (p.v)

Elsewhere Wooldridge (1958, p.31) emphasized that geomorphology should be mainly concerned with the interpretation of landforms *not* the study of processes.

> Compare the change of viewpoint represented by the switch to the quantitative study of processes with the paradigm switch involving the theory of plate tectonics. To what extent do you think that they were due to the same kinds of factors?
>
> Compare and contrast the role which observation and evidence played in each change.

Case Study 4.3: The origin of springs and rivers

Aristotle believed that the universe was composed of five elements. In the sphere below the moon things were subject to generation and decay and were composed of the four elements: earth, water, air and fire. In the sphere above the moon things were ungenerated and indestructable. The heavenly bodies which occupied the sphere above the moon were composed of the fifth element. The four earthly elements had four qualities: cold-dry (earth), hot-dry (fire), cold-humid (water) and hot-humid (air). Aristotle believed that each element could generate the others in a circular way (see Figure 4.4). He used this idea to explain the origin of rainfall and the origin of springs and rivers.

Biswas (1970) considers that, although Aristotle accepted that rainfall could contribute to the flow of rivers and springs, he thought the source of flowing water was largely independent of precipitation. Aristotle argued that since rain is produced when cold air changes to water above the earth, a similar effect must occur within the earth. Aristotle considered that springs and rivers resulted from the continuous conversion of air to water inside the earth. Thus if mountains were large, dense and cold they would catch and retain water so that the rivers they generated would be perennial. If the mountains were small, porous and stony they were likely to lose their store of water quickly and have rivers which flow seasonally.

> Do you find Aristotle's theory about the origin of springs and rivers understandable?
>
> How would you go about evaluating his ideas?

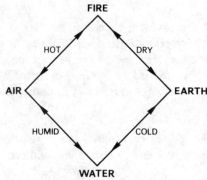

Figure 4.4 Aristotle's concept of elements
Source: after Biswas 1970

More on paradigms

Kuhn's ideas have been widely employed to describe the history of science and of geography in particular. Indeed the Kuhnian sense of the word paradigm has become part of the everyday vocabulary of many scientists. Yet, despite the widespread acceptance of the concept of a paradigm, it is not as straightforward an idea as it seems on first acquaintance.

Both Shapere (1964) and Masterman (1970) have noted the diversity of meanings originally associated with the term paradigm contained in Kuhn's *Structure of Scientific Revolutions.* Shapere, for example, found that the word was used to cover:

> . . . laws, theories models, standards, methods (both theoretical and instrumental), vague intuitions, explicit or implicit metaphysical beliefs (or prejudices). In short, anything that allows science to accomplish anything can be part of (or somehow involved in) a paradigm. (1964, p.385)

Shapere argues that the blanket use of a single term to describe such a range of elements fatally undermines Kuhn's account. He concludes that Kuhn's view is

> . . . made to appear convincing only by inflating the definition of 'paradigm' until that term becomes so vague and ambiguous that it cannot be easily withheld, so general that it cannot easily be applied, so mysterious that it cannot help explain, and so misleading that it is a positive hindrance to the understanding of some central aspects of science . . . (p.393)

Paradigms include everything and so explain nothing. The validity of Kuhn's thesis that shared paradigms are the common factors which guide scientific progress is guaranteed by the very generality of the term. Skolimowski (1974, pp.490–491) has speculated that Kuhn's model might not have become so popular had the term paradigm been given one precise meaning. Stoddart (1977) has made a convincing case that this has occurred in geography.

Kuhn (1977) has acknowledged that the word originally had a variety of meanings. In the postscript to the second edition of his *Structure of Scientific Revolutions* (Kuhn 1970a, 1977; Suppe 1977, pp.136–137) he tried to describe more clearly what he had originally intended. He isolated two central ideas implied by his earlier use of the word. These are the *disciplinary matrix* and *exemplars*.

The disciplinary matrix is the set of shared beliefs and assumptions which account for the consensus in a scientific community. Kuhn uses the idea to describe the intangible factors which produce consensus implied by the earlier idea of a paradigm. Kuhn (1970a) suggests that a disciplinary matrix includes shared models and symbolic generalizations, along with sets of shared values about such things as the character of acceptable theories and solutions. These commonly accepted beliefs are generally acquired through the educational process. In contemporary geomorphology, for example, elements of the disciplinary matrix might be ideas about various sorts of geomorphological process, or some theory, such as plate tectonics, or some set of concepts like 'systems theory'.

In contrast to the ideas represented by the disciplinary matrix, exemplars are the more tangible things which produce consensus in a scientific community. Thus exemplars are the 'concrete and archetypal applications' of the ideas contained in the disciplinary matrix. In geomorphology exemplars would include landscape features regarded as 'typical examples' of particular forms, or which illustrate some underlying process like 'weathering' or 'feedback'. According to Kuhn, the scientist acquires knowledge of the disciplinary matrix by the study of such exemplars. By exposure to these exemplars the scientist learns what kinds of question to ask and what kinds of solution are acceptable.

Replacing the diffuse idea of the paradigm with the more clearly defined ideas of the disciplinary matrix and exemplars, Kuhn suggests that the character of normal science can be more easily understood. He proposes that normal science is carried out by a community bound together by these common elements and is primarily concerned with solving the open-ended puzzles posed by the exemplars and the disciplinary matrix. The most important implication of this view concerns the basis of comparison between alternative conceptual frameworks when the consensus of normal science breaks down. In order to explore this implication we must examine the problem of incommensurability.

Incommensurability and relativism

Kuhn (1970a, p.266) notes that a comparison between two theories demands the availability of a 'common language'. Not, that is, a language in the sense that English is a language, but a kind of neutral vocabulary in which each theory can be discussed and compared. Kuhn suggests that the existence of such a language has been assumed since the seventeenth century and taken to be provided by the neutrality of pure observation. Kuhn denies that such a neutral language exists. Observations are inextricably bound up with theory. In the transition from one theory to another, words change their meaning and conditions of applicability change in subtle ways. The alternative conceptual frameworks in which the theories are contained prevent uncorrupted translations of ideas so the theories cannot be compared. Kuhn describes this problem by saying that theories are *incommensurable*.

According to Kuhn, the ideas of the rationalist break down because of the problem of incommensurability. 'Good reasons' based on logic and observation cannot be used as the basis of theory choice because the words in which ideas are expressed depend on the paradigm or research tradition in which the work is carried out. There is, therefore, no *rational* basis for comparison between them.

Hacking (1983) summarizes the problem of incommensurability as follows:

> ... successive and competing theories within the same domain 'speak different languages'. They cannot be strictly compared to each other nor translated into each other. The languages of different theories are the linguistic counterparts of the different worlds we may inhabit. We can pass from one world or one language to another by a gestalt-switch, but not by any process of understanding. (p.66)

If alternative frameworks and theories are incommensurable, then they interpret facts in different ways. They may even disagree as to what the facts are. They ask different kinds of questions and apply different standards in the evaluation of their solution. What then is the basis of any attempted comparison between them? As the consensus of normal science dissolves and the dominance of one framework declines, how is the scientist to decide which framework he should choose?

Kuhn argues that the choice between alternative frameworks cannot be made by appealing to anything like proof or falsification. The conceptual frameworks represented by different disciplinary matrices simply cannot be compared in this way. On the choice between paradigms Kuhn had earlier argued:

> The substitution of one paradigm for another is not a matter that can be settled entirely by reference to logic or experiment. It is a matter of judgement, an act of subjective choice, an act of faith ... (1970a, p.37)

The same holds true for different disciplinary matrices. The 'good reasons' of the rationalists must be replaced by the ideas of relativism. Scientific judgements are made relative to an accepted set of community standards and norms.

The problem of incommensurability is a difficult one because, according to Hacking (1983), there are at least three ways in which theories can be said to be incommensurable. It is useful to consider each of these in more detail because they illustrate the reasons which lie behind Kuhn's relativism.

The first aspect of incommensurability which Hacking describes is 'topic-incommensurability'. This arises if a new theory simply tackles different problems or uses different concepts from the original. Since the two theories or paradigms do not cover the same subject matter there is no basis for comparison between them. Features of this type of incommensurability are apparently illustrated by the comparison between historical geomorphology represented by Davis and the quantitative, process-response approach. As Case Study 4.2 shows, the switch between them had very little by way of a 'rational basis'. There are no experiments or observations which can be pointed to which refute one and support the other. The quantitative revolution is not usually represented in terms of the rationalist image of science; the change of orientation in the subject is usually described in terms of the sociology of science rather than its logic (cf. Taylor 1975). Such an interpretation is questionable, however, and we will return to the issue in Chapter 7.

The second type of incommensurability is 'dissociation'. This occurs where, after sufficiently long time, or after a sufficiently large number of changes, a theory may become completely unintelligible to a later generation. Although the modern student may find the ideas of Davis unfamiliar, if his writings and those of his contemporaries were studied, the ideas could be easily understood. With the ideas of Aristotle (Case Study 4.3) things would be more difficult. In the attempt to understand his ideas it is not a matter of simply understanding an unknown theory, but also of understanding the way in which Aristotle thought and argued. To do so would require us to dissociate ourselves from the way of thinking of our own time and to reconstruct the intellectual framework of Aristotle. With this kind of dissociation any basis for a rational comparison between different world views is supposed to disappear. Tinkler (1985, p.11–12) gives an additional example of this type of incommensurability from Hutton's writings on the ability of alpine glaciers to erode.

The final type of incommensurability is connected with the meaning of the words in which any comparison between theories is attempted. Meaning-incommensurability comes about because many consider that the meaning of scientific terms can only be understood in the context of a theory. With any change in theory the meaning of words will change. Thus once again there never can be a rational basis for the comparison between paradigms. If this type of incommensurability occurs, the choice between theories is always subjective.

If theories are incommensurable, then there is no rational basis for choosing between them. Choice is subjective, and can only be understood by the appeal to psychological, sociological or even political forces. On the issue of paradigm choice Kuhn concludes:

> As in political revolutions . . . there is no higher standard than the assent of the relevant community. (1970a, p.94)

On theory choice

In order to avoid the charge that he has reduced the basis of theory choice to a matter of 'mob psychology', Kuhn has attempted to make the decision criteria more explicit.

He suggests, for example, that there are at least five characteristics that might lead one to prefer one theory over another (1977 pp.321–339). These are:

1. A theory should be *accurate* within its own terms of reference.
2. A theory should be *consistent* both internally and with other accepted theories.
3. A theory should be *broad in scope*, with implications beyond its immediate terms of reference.
4. A theory should be *simple*.
5. A theory should be *fruitful*, that is, it should lead to new findings.

On the face of things these criteria seem easy enough to employ. In the case of the theory of plate tectonics, for example (Case Study 4.1), the theory is accurate. The rates of plate movement currently detected are of the correct order to explain the separation of continental blocks required by the period of geological time which has elapsed since their supposed fracture. The theory is, moreover, consistent with the kinds of process supposed to be operating within the earth. The theory is simple, yet has implications beyond its immediate terms of reference. It can explain the previously problematic observations relating to palaeoclimatic and palaeobio-geographical distributions which arise on a static earth. Finally, the theory would appear to be quite fruitful. It poses many puzzles which earth scientists can pursue and their success is to a large extent guaranteed by the general framework. What, however, is the status of these characteristics which might suggest that one theory is better than another?

At one time Kuhn (e.g. 1970a, p.185) appears to suggest that the five criteria described above are part of the disciplinary matrix. Elsewhere (1977, p.322) he seems to suggest that they are independent of the framework provided by any one disciplinary matrix. They are the 'shared basis of theory comparison' (1977, p.321). These ideas clearly conflict. Either the bases of comparison are part of the disciplinary matrix and theories are incommensurable, or the criteria allow comparison to be made between theories across different frameworks. If this is so, then Kuhn's thesis of relativism cannot be upheld.

In his later writing, Kuhn (1977) appears to be arguing that there is some shared basis for theory choice and thus appears to be taking up a position closer to the rationalists. Kuhn continues to differ from the rationalists in one important respect, however. He considers that, although there might be a shared basis of comparison, these criteria can have no justification. As Newton-Smith (1981, p.121) observes, according to Kuhn, all progress in science is to be done externally, not internally. It must:

> ... in the final analysis, be psychological or sociological. It must, that is, be a des-cription of a value system, an ideology, together with an analysis of the institutions through which that system is transmitted and enforced. Knowing what scientists value, we may hope to understand what problems they will undertake and what

choices they will make in particular circumstances of conflict. I doubt that there is another sort of answer to be found. (Kuhn 1970b, p.21)

According to Kuhn there is no way in which a decision can be grounded solely on evidence. Scientists may agree that consistency or simplicity are characteristics of theories which aid decisions, but disagree about the way in which these features figure in a particular dispute. In addition there may be disagreement as to the relative importance to be attached to each characteristic. Thus, while rules of comparison may be employed, so that there is a pattern to the decisions made, one cannot assume that the decisions are rationally justified in the same way as writers such as Popper argue. A significant element of Kuhn's relativism thus remains.

Kuhn is not the only one to view science in terms of relativism. Feyerabend, for example, has also argued that the development of science can only be understood in terms of the social and intellectual framework of the community of scientists. Before we can make any final assessment of relativism we must move on to consider some reactions to the view of science proposed by Lakatos and Feyerabend. Try Exercise 4.3.

CHAPTER 5:
LAKATOS AND THE NATURE OF RESEARCH PROGRAMMES

'All theories are born refuted'

Introduction

The ideas of Kuhn and Popper stand out in stark opposition to each other. On the one hand, Kuhn maintains that scientific judgement has little by way of a rational basis. There is a pattern to the judgements which scientists make, and this pattern is shaped by accepted standards and values, but there is no wholly rational justification for the standards and values themselves. Decisions are made relative to some conceptual framework. Popper, on the other hand, argues that science is a rational activity. It can provide 'good reasons' for the judgements it makes. The rational basis of judgements is provided by the critical investigation of theories and the ideas of falsification and falsifiability. One interpretation of Latakos's work is that it represents an attempt to resolve these conflicting views and to produce a methodology that can combine the strengths of each of them. In the conclusion to one of the major statements of his position he writes:

> . . . I have tried to develop his [Popper's] model a step further. I think this small development is sufficient to escape Kuhn's strictures. (1970, p.179)

Let us see what kind of modification he had in mind.

The methodology of scientific research programmes

Lakatos points out that if the principle of falsification is used as a guide to judgement then 'all theories are born refuted'. No theory agrees with all known facts yet this is not always grounds for theory rejection. Despite its excellent logical credentials, the principle of falsification, if applied strictly, would kill off theories too quickly. If we attempt to operate the principle rigidly then any kind of growth in scientific knowledge would be prevented. For, according to Lakatos, scientific knowledge does grow, in the sense that over time the scientist can come to know more about the world than was known in the past. As Hacking (1981) notes, this is one of the key assumptions which Lakatos makes. How can such growth occur despite the fallibility of most theories? What is it that keeps theories alive and fosters their development?

For Lakatos it is not the fate of individual theories which shapes the growth of knowledge but much larger frameworks of thought. He refers to such frameworks as 'research programmes'. In many respects his idea of a research programme is similar to Kuhn's idea of a paradigm although, in order to keep the different views in focus, it is not worth developing the analogy too far. Lakatos has several very specific kinds of thing in mind, beyond simply a 'world view' or 'exemplar', when he refers to a research programme. Research programmes guide scientific activity because they contain certain key elements not explicitly recognized in Kuhn's idea of a paradigm.

According to Lakatos, research programmes contain two distinct elements which guide research in quite distinct ways. One element (the negative heuristic) tells the scientist which paths of research he should avoid. The other (the positive heuristic) suggests which pathways should be pursued. By the word heuristic Lakatos means 'a way of doing things', that is, a methodology. Thus the negative heuristic is simply the decision that certain key assumptions of the research programme will go unquestioned. These assumptions represent the so-called hard core of the research programme. They are those ideas which are taken, by a matter of convention, to be irrefutable or at least beyond doubt. The negative heuristic directs research away from a critical examination of these key ideas. Lakatos considers that the hard core of a research programme is protected by a belt of 'auxiliary hypotheses' and other ideas about which the scientific community are prepared to take a more flexible attitude. The positive heuristic is simply the set of informal guide-lines which suggest how this protective belt might be investigated, and where modifications might be made in order to resolve any anomalies thrown up by the programme of research. It is within this protective belt that refutation can happily take place, but while these subsidiary ideas may be replaced, they are modified in such a way that the hard core is maintained. Lakatos summarizes his ideas about the way in which scientific knowledge grows as follows:

> The basic unit of appraisal must not be an isolated theory or conjunction of theories but rather a *research programme* with a conventionally accepted (and thus by provisional decision 'irrefutable') *hard core* and with a *positive heuristic* which defines problems, outlines the construction of a belt of auxiliary hypotheses, foresees anomalies and turns them victoriously into examples, all according to a preconceived plan. The scientist lists anomalies, but so long as his research programme sustains its momentum, he may freely put them aside. *It is primarily the positive heuristic of his programme, not the anomalies, which dictate the choice of his problems.* (1981, p.116)

This extract from Lakatos is useful because it begins to bring out the differences between his model and those of Popper and Kuhn. Lakatos is sceptical of the naive application of Popper's falsificationism. If the growth of scientific knowledge depends upon the fate of individual theories, then most theories would be rapidly discarded as anomalies accumulate. As Chalmers (1982, p.85) notes, when the Popperian is faced with a refutation, the fact that any part of the theory and its conceptual background may be responsible for the falsification poses a serious difficulty. For the Popperian the inability to locate the source of the problem results in 'unmethodical chaos'. Chalmers suggests that the research programmes which

Lakatos has in mind are sufficiently well structured to avoid such consequences. Order is maintained by the decision to accept the hard core as being above challenge. The problem of apparent refutation is solved by the modification of the protective belt of ideas.

Why are apparently refuted theories maintained? According to Lakatos it is because they are part of larger programmes of research. These are the units of judgement which finally seal the fate of individual theories. Providing the scientist is prepared to accept the basic assumptions of a particular research programme, then anomalies can be tolerated. They are a stimulus to research and are responsible for the development of the research programme.

Research programmes in physical geography

In the same way that geographers have found it easy to recognize paradigms in geography, so distinct research programmes, with their positive and negative heuristics can also be discovered (cf. Wheeler 1982). The following two case studies illustrate some aspects of Lakatos's ideas. They will be discussed in the sections which follow.

Case Study 5.1: Plate tectonics and the fossil record

The theory of plate tectonics could be described in terms of Lakatos's idea of a research programme. In many ways it represents a more complex set of ideas and assumptions than one finds in many theories. On inspection it appears to include elements which correspond to Lakatos's ideas of a hard core and a protective belt of auxiliary hypotheses.

For example, as part of the hard core of the research programme, one might recognize the basic assumption of a mobile earth. That is, the idea that the surface of the earth consists of rigid, but nevertheless mobile plates (see Case Study 4.1). Although details about the location and character of the plates varies, this basic idea is applied in a variety of situations. Although individual theories based on this idea may fail, such experience is not taken to refute the basic assumptions of plate tectonics.

Consider, for example, the problem posed by the biogeographical distribution of the three late Cretaceous dinosaurs shown in Figure 5.1. Hallam (1972) notes that all three fossil formations are present in South America and India. One genus, *Titanosaurus*, is present in both Europe and Africa, and another, *Laplatosaurus*, is also found on Madagascar.

Unless the fossil identifications are mistaken, these biogeographical distributions imply that at one time the land masses must have been connected in order that the dinosaurs could have spread so widely. The fragmented pattern of 'disjunct endemism' that we currently observe in the fossil record is the result of plate movement, but these biogeographical distributions pose a problem. The fossil record shows that the dinosaurs were only present in the late Cretaceous. The geological

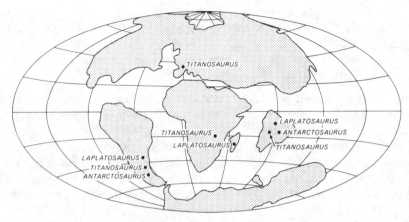

Figure 5.1 The dilemma posed by the distribution of three genera of late Cretaceous dinosaurs. The biogeographical distributions imply that the continents were joined in the late Cretaceous when other geological evidence suggests that the continental blocks were already well separated by this time.
Source: after Hallam 1972

evidence used by Hallam (1972) to reconstruct the position of the continents at this time suggests that the continents were already separated. As Figure 5.1 shows, in late Cretaceous times India had already split from the African-Arabian block, continuing a trend which had begun some 100M years earlier. According to the biogeographical distribution the continental masses could not have been severed at so early a date.

What other explanations for the anomaly can you think of? How would you test these ideas?

Do you consider that the problematical distribution of the late Cretaceous dinosaurs refutes the theory of plate tectonics?

Case Study 5.2: Landscape sensitivity and change
The account of Brunsden and Thornes (1979) can be viewed as a summary of the major features of the process-response research programme of contemporary geomorphology. In their essay they aimed to:

... review some of the more general concepts which have been generated by the process-form studies and to consider whether they yield a coherent conceptual basis for studies of long-term landform evolution. (p.464)

Their account is especially interesting since they state that the dynamic basis of modern geomorphology has depended upon the adoption of 'the methodology of seeks to penetrate 'behind' the external appearances of phenomena to the 'essences of mechanisms which necessitate them'.

Brunsden and Thornes state the basic concepts generated by the process-form studies as a series of propositions:

1. For any set of environmental conditions, through the operation of a constant set of processes, there will be a tendency over time to produce a set of characteristic landforms (see Figure 5.2).
2. Geomorphological systems are continually subject to perturbations which may arise from changes in the environmental conditions of the system or from structural instabilities within. These may or may not lead to a marked unsteadiness or transient behaviour of the system over a period of 10^2–10^5 years (see Figure 5.3).
3. The response to perturbing displacement away from equilibrium is likely to be temporarily and spatially complex and may lead to a considerable diversity of landform (see Figure 5.4).
4. Landscape stability is a function of the temporal and spatial distributions of the resisting and disturbing forces and may be described by the landscape change safety factor here considered to be the ratio of the magnitude of barriers to change to the magnitude of disturbing forces (see Figure 5.5).

Consider these propositions and for each of them try to give an example of a geomorphological situation which illustrates the idea.

Consider the propositions carefully and try to identify what kinds of observation would lead you to think that the propositions were false.

To what extent would you regard these propositions as forming part of the 'hard core' of modern geomorphology? Give an example of work from the 'protective belt of auxiliary hypotheses'.

realist science' (p.463). According to Chorley (1978, p.2), realist theory in geomorphology views explanation as more than prediction based on observation. It

The choice between research programmes

There are research traditions in physical geography and the natural sciences which have some of the characteristics of the kind of research programme described by Lakatos. As Case Study 5.1 illustrates, the theory of plate tectonics has ideas which might be described as representing a hard core of accepted assumptions, and these support work which apparently only deals with the modifiable belt of auxiliary hypotheses. Despite the apparent conflict between the biogeographical and the geological observations, the anomalous biogeographic patterns described in the case study are not likely to be taken as a refutation of the theory of plate tectonics. Instead it is regarded as the kind of difficulty which one would normally expect to encounter in such a research programme and which could be resolved by a modification of the auxiliary hypotheses built on the basic assumption of a mobile earth. The anomalous

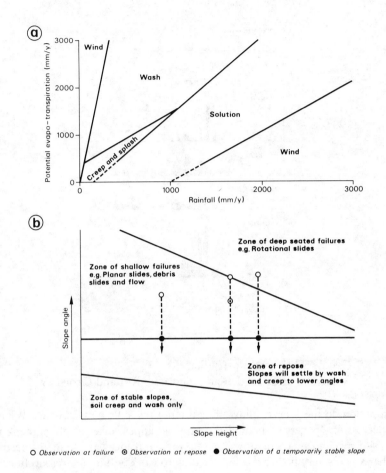

Figure 5.2 Process-domains
Source: after Brunsden and Thornes 1979

distribution could be explained by the suggestion that movements of plates have occurred but not perhaps in the way originally envisaged. Alternatively, perhaps the factors controlling the biogeographical distribution of dinosaurs are not as we imagine. The central concepts of plate tectonics are not, however, to be questioned. Such a strategy would seem sensible. But then the question arises as to whether we should always act in this way. The kind of problems to which an uncritical approach might lead are illustrated in Case Study 5.2.

The ideas described by Brunsden and Thornes (1979) in Case Study 5.2 are presented as some of the basic principles of contemporary geomorphology. It is interesting to note, however, that they are not presented in a form which is in any way testable. They contain phrases such as 'may or may not' which insulate them from any kind of refutation. If there is no way in which the principles can be tested, then

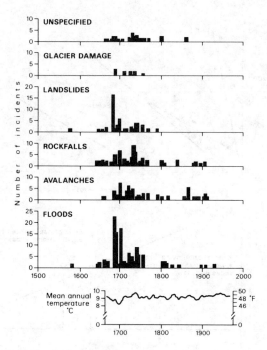

Figure 5.3 Geomorphological unsteadiness
Source: after Brunsden and Thornes 1979 and Grove 1972

what function do they serve? Since they are not testable, how can we tell whether they successfully 'penetrate behind the external appearances of phenomena to the essences of mechanisms which necessitate them'? If ideas are treated uncritically, then there seems very little basis for deciding to work within one research tradition rather than some other. How is the choice between research programmes to be made?

The concept of a research programme with its ideas of a hard core and a protective belt of modifiable ideas represents an attempt to describe a more conservative methodology than that provided by Popper. Lakatos also argues that it gives a better account of the history of science, in which there are many examples of people persisting with a particular line of research despite the acknowledged existence of data which run counter to its predictions. Although Kuhn regards persistence with such particular world views as largely irrational, and explains it in terms of psychology and sociology, Lakatos seeks to retain a basis in reasoned argument. The way in which reason shapes development may be illustrated by Lakatos's account of the way in which one research programme may eventually replace another.

Lakatos describes research programmes as either 'progressive' or 'degenerating' according to whether they lead to the discovery of new things. A research programme is progressing as long as its protective belt of hypotheses can be modified to take account of new data, not in an ad hoc way, but in a manner which can lead to the pre-

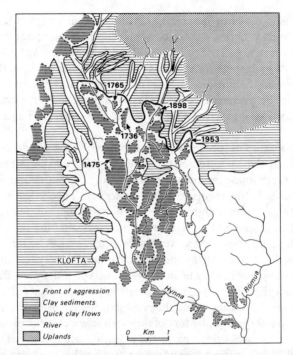

Figure 5.4 The propagation of a geomorphological change in the landscape as a diffusing wave
Source: after Brunsden and Thornes 1979

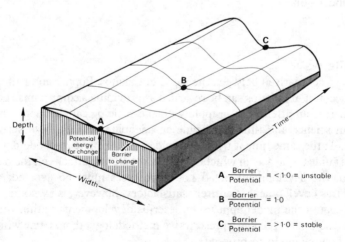

Figure 5.5 Model showing how, through time, geomorphological systems become more entrenched in furrows in the 'space–time manifold', such that the barriers to change become greater and more energy is required to change from one equilibrium to another (after Brunsden and Thornes 1979)

diction of novel facts. Lakatos describes this situation in terms of a 'progressive problem-shift'. If such progress cannot be achieved, then the research programme stagnates (i.e. there is a 'negative problem-shift) and scientists may be led to adopt a rival programme. At each stage the scientist makes a judgement about the value of a particular research programme on the basis of its potential compared to any rivals. For Kuhn the choice between rival paradigms is a matter of faith or 'mob psychology'. For Lakatos, the decision is based on a judgement of performance. The decision is not an irrational act.

Many of the properties of 'progressive' research programmes can be seen in the examples outlined in the case studies presented in this and the last chapter. The mobilist research programme produced first the theory of continental drift, then, as new data accumulated, the theory of plate tectonics. Theoretical developments anticipated and stimulated new empirical observations. A more dynamic programme of research could be sustained by the ideas of mobilism than by the older stabilist view. As a consequence it is hardly surprising that a major switch of allegiance occurred in the earth sciences during the 1960s. In a similar way one can argue that the process-response approach became more widely accepted than the older historical approach to geomorphology as its greater potential was appreciated.

The examples of mobilist theory and the process-response approach illustrate that one does not have to interpret the history of the earth sciences in terms of sociological and psychological processes operating in the scientific community. To admit that changes in outlook have occurred does not mean that one has to accept Kuhn's interpretation of history. With Lakatos's and the benefit of hindsight, the developments can look quite rational. Does Lakatos therefore offer a viable compromise between Popper and Kuhn?

A middle way?

In making a comparison between the views of Lakatos, Popper and Kuhn, it is clear that the account of Lakatos can be regarded as providing a compromise between two extremes. It is an attempt to grapple with the sociological elements which clearly operate in science and thus to provide an adequate *description* of scientific communities. At the same time it tries to provide a prescriptive methodology which can serve as a guide to the way in which scientists should act: attacks on the hard core are forbidden, the development of ad hoc hypotheses must be avoided, and so on. Whether this development is entirely satisfactory, however, is by no means certain. Merely because the model appears to describe developments within the discipline should it also be accepted as a prescriptive methodology, that is, one which tells the scientist how he ought to proceed?

Although the methodology suggested by Lakatos appears to resolve many of the problems arising from the conflict between rationalism and relativism, it has itself been criticized. A major difficulty stems from the fact that it is impossible to decide which research programmes are progressing and which degenerating until after the event. Lakatos provides no forward-looking assessments of competing scientific theories:

> He can at best look back and say why, on his criteria, this research programme was progressive and why another was not. (Hacking 1983, p.121)

Feyerabend (1970) suggests that the critical standards which Lakatos employs (i.e. the positive and negative heuristics) provide an interval for hesitation in the evaluation of theories. He argues that such standards only have practical force if they are combined with some time limit, but if this is done, then all the arguments against the falsificationist approach reappear:

> Thus the standards which Lakatos wants to defend are either *vacuous* – one does not know when to apply them – or they can be *criticised* on grounds very similar to those which gave rise to them in the first place. (p.215)

In these circumstances, Feyerabend concludes that we can do one of two things. On the one hand, we can reject the rationalist view that there are permanent standards for the evaluation of theories, which remain in force throughout the history of science and which apply to every period of scientific development. Alternatively, we can:

> ... retain such standards as a *verbal ornament*, as a memorial to happier times when it was still thought possible to run a complex and often catastrophic business like science by following a few simple and 'rational' rules. (ibid.)

For Feyerabend, one of the severest critics of rationalism, Lakatos's ideas represent no advance over Popper's at all. The issue remains of how the scientist can ever find 'good reasons' for preferring any one theory to another. Before we can resolve the issue, we must look at the arguments which Feyerabend mounts against rationalism. Does he succeed in overturning the rational image of science?

CHAPTER 6:
FEYERABEND ON METHOD

'Anything goes'

Introduction

Many people consider science to have a special place amongst human activities. It is often represented as being more logical, objective and capable of providing deeper insights into our world than any other undertaking. The products of science are, it is often suggested, of a more practical and lasting kind. Those who probe this image and who attempt to reconstruct the rationale which has lain behind scientific progress in their chosen field, sometimes find that the picture is more confused. The logic on which progress is supposed to be based is frequently less than clear. Scientists often conspicuously lack the objectivity they are supposed to possess. And many of the products of science tend to threaten society rather than improve it. Why should the image of science be so different from its reality?

Perhaps one explanation for the gap between the image of science and its reality is due to our understanding of its methods. Alternatively, perhaps there is no method. Science might genuinely be an anarchistic undertaking. Up to this point we have assumed that the fault lies with an understanding of method. By our summaries of the classical model and the ideas of Popper, Kuhn and Lakatos, we have described the various attempts to expose the methodology which underlies scientific activity. What of the alternative pathway, the one which questions whether any sort of pattern can be detected at all? In order to explore this we must turn to the ideas of Feyerabend.

Feyerabend (1975, 1978) has rejected the idea that anything like a scientific method can be identified. He turns away from any attempt to identify yet another scientific methodology and attacks *all* methodologies. Feyerabend outlines his aim as follows:

> I want to defend society . . . from all ideologies, science included. All ideologies must be seen in perspective. One must not take them too seriously. One must read them like fairytales which have lots of interesting things to say but which also contain wicked lies, or like ethical prescriptions which may be useful rules of thumb which are deadly when followed to the letter. (1981, p.156)

By 'ideology' Feyerabend means a particular viewpoint. His goal is to undermine the idea that science has any special place in the world of ideas and in society. In this chapter we will examine the challenge he offers to the rational image of science.

The myth of science

The myth that science holds an exceptional place in society is, Feyerabend (1975, 1978, and 1981, p.158) claims, based upon two assumptions:

1. that science is the activity which has discovered the *correct method* of achieving results;
2. that there are many *results* which demonstrate the excellence of the method.

With regard to the first point Feyerabend suggests that the idea that there is such a thing as a scientific method, consisting of firm and binding principles, meets with considerable difficulties when confronted by the results of historical investigation. If we look into the history of science:

> We find . . . there is not a single rule, however plausible, and however firmly grounded . . . that is not violated at some time or another. It becomes evident that such violations are not accidental events, they are not results of insufficient knowledge or of inattention which might be avoided. On the contrary, we see they are necessary for progress. (1975, p.23)

Feyerabend's attack on method takes the form of showing that, whatever rules one tries to identify (verification, falsification etc.), one can find historical cases where they have been broken, and where scientific progress has, nevertheless, occurred. Science is an anarchistic undertaking. The only methodological principle that can be defended in all circumstances is the principle that states that *anything goes* (1975, p.28; 1978, p.39).

Feyerabend attacks the idea that correctness of scientific method is borne out by the concrete results it has provided by arguing that, while the achievements of science are manifest, so are the achievements of many other ideologies. Science is not unique in the understanding it provides. It can have an influence on society but only to the same extent as any other pressure group. Science is simply one of the many ideologies which propel society and should be treated as such. In support of his argument Feyerabend (1975, p.49) cites the case of Copernicus.

Copernicus reintroduced the idea that the earth moves, despite the fact that it conflicted with accepted contemporary ideas. He maintained it, despite all sound rules of scientific practice. The theory conflicted with accepted facts and even its predictions were in error. Copernicus took an idea regarded by most to have been thrown on the 'rubbish heap of history' and forged it 'into a weapon for the defeat of its defeaters'. Progress did not, in other words, occur by any rational process. Instead it occurred more as a result of individual motivation and superior propaganda. Although the results are now part of 'knowledge', they were not fashioned by any superior scientific method. A similar point might be made about the influence of Davis's cycle of erosion (see Bishop 1980, p.314).

As a further example of the way in which ideas from outside science can, nevertheless, propel society and lead to scientific progress, Feyerabend cites the case of medicine in Communist China. Following the Cultural Revolution in 1966, traditional herbal medicines and acupuncture were ordered back into universities and hospitals despite the fact that they were regarded with scepticism by the scientific community:

> ... there was an outcry all over the world that science would be ruined in China. The very opposite occurred: Chinese science advanced and Western medicine learned from it. (1981, p.162).

Feyerabend argues that there are many examples where we can see great scientific advances occurring due to influences outside science. Ideas are made to prevail in the face of the most basic and most rational set of methodological rules. Feyerabend concludes that the results of science are hardly ever achieved by science alone.

Feyerabend and scientific practice

Feyerabend attacks science on two fronts. On the one hand he questions the existence of scientific method. On the other he questions the supposed superiority of scientific knowledge over that provided by other kinds of undertaking. In order to illustrate the kinds of evidence he puts forward to support his views and the possible relevance of these ideas to geography, consider the following case studies. You might also try Exercise 6.1.

Case Study 6.1: The triumph of uniformitarianism

The idea that the features of the earth's surface and the underlying geological strata have been formed over long periods of geological time, by the kinds of process that we currently see around us, is often regarded as a basic assumption of contemporary earth science. Indeed, so basic is this idea that it is often left unstated. Yet when the idea was put forward by Hutton in the eighteenth century, it was a view which differed radically from the prevailing belief that earth history could only be understood in terms of the influence of a series of cataclysmic events. Gillispie (1959) notes that in the Britain of the 1820s, the popular conception of geology was practically synonymous with the doctrine of catastrophies. In particular, the Biblical Flood was regarded as the primary geological event. How did the idea of 'uniformitarianism' come to replace the entrenched and established views of the 'catastrophists'? What role did scientific method play in the matter as compared to influences outside science?

A major influence on the shaping of catastrophist ideas was Baron Cuvier. In the early 1800s Cuvier aroused much interest by his study of the bones of dinosaurs, and the implications which could be drawn from their study for geological history. He was able to show that, in areas presently inhabited, there was evidence that animals very different from contemporary forms had existed and that earlier episodes had been ended by invasions of the sea or floods. In his *Recherches sur les Ossemens Fossiles* Cuvier wrote:

> The changes in height of the waters did not consist simply of a more or less gradual and universal retreat. There were successive uprisings and withdrawals of which, however, the final result was a general subsidence of sea level.
>
> But it is also extremely important to notice that these repeated inroads and retreats were by no means gradual. On the contrary, the majority of the cataclysms that produced them were sudden. This is particularly easy to demonstrate for the last one . . . It also left in the northern countries the bodies of great quadrupeds, encased in ice and preserved with their skin, hair, and flesh down to our own times. If they had not been frozen as soon as killed, putrefaction would have decomposed their carcasses . . . The dislocations, shiftings and over-turning of the older strata leave no doubt that sudden and violent causes produced the formations we observe, and similarly the violence of the movements which the seas went through is still attested by the accumulations of debris and of rounded pebbles which in many places lie between solid beds of rock. (Translation in Gillispie 1959, pp.99–100)

Cuvier's ideas were championed and developed in Britain by Buckland and his associates. As Chorley et al. (1964) note, the early years of the nineteenth century were a period of active field study, and workers such as Buckland sought to reconcile the accumulating evidence of the geological record with the Biblical story. These workers, often referred to as the 'diluvial' school, argued that Noah's Flood as described in the book of Moses, was the major, most recent geological event to have shaped the surface of the earth.

In the same way that catastrophist ideas had their champions, so the uniformitarian ideas of Hutton and his associates were also actively propagandized. Although he was at one time a catastrophist, Charles Lyell later became a key figure in promoting the idea of uniformitarianism. In support of his ideas Lyell argued for the effectiveness of river erosion as a process shaping the earth's surface:

> We have but to consider the mountains as formed by the hollowing out of the valleys, and the valleys as hollowed out by the attrition of hard materials coming from the mountains . . . (Hutton 1795, II, p.401, quoted in Chorley et al. 1964, p.40)
>
> Our solid earth is everywhere wasted, where exposed to the day. The summits of the mountains are necessarily degraded. The solid weighty materials are everywhere urged through valleys, by the force of running water. (Hutton 1795, II, p.561, quoted in Chorley et al. 1964, p.40)

As a mark of the emphasis Lyell and his associates placed upon the work of fluvial erosion, they were often referred to as the 'fluvialist' school.

The kind of evidence put forward by the diluvialists to support the idea of the Biblical Flood included the large deposits of ill-sorted sands and gravels, known as 'drift', found to be plastered onto the landscape in many parts of Britain. Frequently, the material of these deposits included erratics, that is rocks different in character from the underlying strata. Elsewhere it was pointed out that rivers were much smaller than the valleys which they presently occupied, and river and marine terraces must indicate a recent and rapid drop in water level. Buckland based his arguments about the Flood on the observation of the bones of animals buried in the mud, silt and clay deposits of caves. His associate Conybeare also pointed out the difficulties of explaining the formation of river valleys by simple erosion in cases, such as the Thames, where the river flows from a clay vale through the upstanding chalk mass of the Chilterns. In the face of such evidence how was the hypothesis of the Flood eventually overturned?

Some key observations which led to the abandonment of the diluvial theory were provided by Lyell and Murchison as a result of their work in Central France. The landscapes of the Auvergne, Velay and Vivarais showed that successive volcanic lava flows had flown out over different rock surfaces 'fossilizing' the landscape beneath them. In a number of cases valleys could be observed beneath lava flows which were themselves below the drift deposits supposed to have been deposited by the Flood (see Figure 6.1). Valleys, it could be observed, had been formed without the intervention of the Flood. Lyell writes of a meeting where the ideas of the diluvialist and fluvialist were discussed, as follows:

A splendid meeting of the Geological Society last night, Sedgwick in the chair, Conybeare's paper on the Thames, directed against Messrs Lyell and Murchison's former paper, was read in part. Buckland present to defend the 'Diluvialists' . . . Murchison and I fought stoutly, and Buckland was very piano. Conybeare's

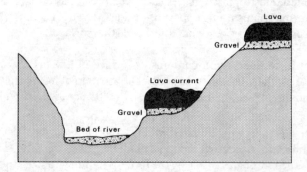

Figure 6.1 The relationship between fluvial and volcanic deposits in the Auvergne, France
Source: after Lyell 1833

memoir is not strong by any means. He admits three deluges before the Noachian and Buckland adds God knows how many more besides; so we have driven them out of the Mosaic record fairly. (Bonney 1895, p.42, quoted in Chorley et al. 1964, p.129)

Buckland and many of the other diluvialists eventually came to reject the Flood hypothesis. In reading of such events it is easy to gain the impression that the matter was decided by logic alone. The diluvial theory was refuted by observations with which it is inconsistent. Yet such stories are perhaps only viable because we can look back and select just those events and extracts that we want. In fact the situation was more complex than the previous account might suggest. The diluvial theory was not replaced by some superior hypothesis which itself could not be refuted. In fact uniformitarianism was itself contradicted by many field observations. At the time of the debate discussed above, the idea of glacial action had not been considered, and so the drift deposits were not explained away by any appeal to river action. In addition, the uniformitarians had no coherent concept of geological uplift to explain such things as the origin of mountains. While they dismissed the intervention of catastrophes, they used evidence supplied by periodic volcanic eruption to argue for the constant and continual action of river erosion to explain the development of the features of the earth's surface, and so on. The point on which the uniformitarians and the catastrophists appear mainly to have differed concerned whether Divine intervention was the primary factor shaping the earth's surface. They by no means ruled out the possibility of large-scale geological events having occurred in the past.

To what extent do you consider that the triumph of the fluvial theory over that of diluvialists can be explained by reference to logic, or merely to more effective propaganda on the part of its protagonists? Can the history of the matter be explained by reference to the operation of anything like a rational scientific method?

Case Study 6.2: The demigod's dilemma

Many social scientists have rejected the methods which are supposed to operate in the natural sciences as providing a model for their discipline. As an example of one such attack on science we will consider the address given to the Association of American Geographers by Zelinsky (1974). In many respects the attack which Zelinsky mounts parallels that of Feyerabend.

Zelinsky, like Feyerabend, observes that science has become the dominant religion of the twentieth century, replacing the traditional faiths of the supernatural:

We have come full circle from a supernatural God-drenched view of life, by way of the fateful Copernican revolution ... to a professedly objective, rational, this-worldly, man-centered theology. In this latter-day dispensation, the supreme, unchallengable doctrine is that of the Scientific Method and its sacred covenant to redeem us all through total and perfect knowledge. (p.125)

He argues that this new religion suffers from a number of serious flaws which render it impotent in the face of many of the problems which concern mankind today.
Science, according to Zelinsky (pp.130–134), is based upon five major axioms:

1. the principle of causality
2. the view that all questions are soluble
3. the belief that science leads to perfect knowledge
4. the universal validity of scientific results
5. the assumption of the objectivity of science

Each of these assumptions, he claims, can be shown to be false. The notion of causality is undermined by the discovery of indeterminate systems; the solubility of all questions is undermined by the evidently intractable social, economic and political issues which face mankind; the acquisition of perfect knowledge is refuted by the repeated overthrow of theory. Finally the objectivity of science is to be doubted once the social context in which science operates is considered:

> Let us face some awkward facts: the choice of respectable or rewarding fields of research is a social decision; deciding whether a given scientific question is scientifically interesting is, again, a social decision; criteria for accepting evidence, methods of investigation, or acceptable proofs and disproofs are arrived at through social consensus, not by reference to external standards graven upon tablets of gold. Since rules may be set by self-perpetuating elites with scant regard for more basic social and intellectual values, some quite idiosyncratic spirals of research may be spawned, so long as they do not run counter to the socially dominant ideologies. More often, however, the social, economic and class interests of the statesmen of science are such, and the interlocking directorates of the scientific, industrial, financial and military communities are so intimately conjoined, that few overt cues are required or needed to make the thrust of either pure or applied research supportive of the status quo and the further accumulation of power and wealth by the already powerful and wealthy. (p.134)

Zelinsky suggests that the flaws in natural science would not be so serious if it were not for the pressure of man's populations on the resources of the biosphere. Such problems go beyond the bounds of the natural sciences, and have implications for the social sciences. However, by adopting the natural science model, social science is poorly placed to make any useful contribution. This is the demigod's dilemma. To what extent would you agree with Zelinsky?

> Do you agree with Zelinsky that the scientific model is flawed? What particular model of science does Zelinsky appear to be attacking?
> Consider any man–environment issue (e.g. acid rain, desertification, industrial

pollution) and try to isolate which elements of the problem are amenable to scientific analysis and which are not.

If you are able to isolate aspects of your environmental issue that do not seem amenable to scientific analysis, do you feel that there is sufficient ground to reject the scientific approach entirely?

Against method

The strategy which Feyerabend uses in his attack is to undermine our confidence in the idea of a rational scientific method by suggesting a set of alternative but equally plausible methodological rules, and then showing that these have been used in science with success. Clearly, such a line of argument will only be effective if he is able to identify the rules which philosophers and scientists assume to have guided rational decision making. Thus, in order to appreciate Feyerabend's argument, it is important to begin by looking at the methodological rules which Feyerabend attacks.

Feyerabend suggests that the decision between competing theories is generally assumed to be guided by two major criteria:

1. that any newly introduced hypothesis must agree with currently accepted theories;
2. that any newly introduced hypothesis must agree with all existing facts, observations and experimental results.

In response to both of these rules, he suggests that it is plausible to act in the opposite way and that this strategy is equally likely to lead to success. Thus his two 'counterrules' are simply that one should attempt to introduce and develop hypotheses which are *inconsistent* with (a) well-established theories, and (b) well-established facts. With such rules, Feyerabend advises the scientist to proceed 'counterinductively'.

In order to show that the first counterrule is plausible Feyerabend (1975, pp.30–31) suggests that evidence which might eventually refute a theory will only emerge with the help of an incompatible alternative. Any scientist who wants to understand the strengths and weaknesses of a theory ought to seek alternative views. Thus the scientist ought to allow, and indeed encourage, the proliferation of mutually inconsistent theories rather than seek a single, consistent view. The demand for consistency is effective only in preserving existing theories rather than producing better ones. Feyerabend suggests that it is plausible for the scientist to adopt a *pluralistic methodology* since 'there is no idea, however ancient or absurd that is not incapable of improving our knowledge' (1975, p.47). The scientist must compare ideas with ideas rather than with experience, and must try to improve rather than discard the ideas which have apparently failed.

To a large extent these views about theory consistency are supported by the case of

the debate between the fluvialists and the diluvialists and the subsequent developments in geomorphological thinking (see Case Study 6.1). If we consider the status of
the uniformitarianism principle that 'the present is the key to the past', as it is currently
accepted in the earth sciences, then it soon becomes clear that it does not mean the
same as when it was first put forward. Although catastrophism has been 'refuted' the
recognition of cataclysmic events is commonplace in geomorphology (Dury 1980),
and the appeal to such events is not taken to be in any sort of conflict with uniformitarianism. A situation which illustrates this point is that of the Scablands of the
Columbia Plateau in Washington, USA.

The loess-covered lava plateaus of eastern Washington are deeply dissected by a
network of valleys known as the Scablands (see Figure 6.2). The valley of the Clark
Fork River, for example, is a deep trough scoured free of soil up to 300m above the
valley floor. Huge gravel benches lie across embayments in the valley side with their
crests more than 150m above the valley bottom. Davis (1921) suggested that such
features were produced by glacial scour and morainic deposition. An alternative
theory has been proposed by Bretz (1923), who explained the Scablands in terms of
the Spokane Flood, the sudden and catastrophic drainage of the pro-glacial Lake
Missoula. As Clayton (1970) notes, while many have been unable to accept Bretz's
theory and have suggested alternative explanations, the only hypothesis which
accounts for all of the field evidence is that of the catastrophic drainage of Lake
Missoula. Pardee (1910) suggested that during the cataclysmic drainage episode at

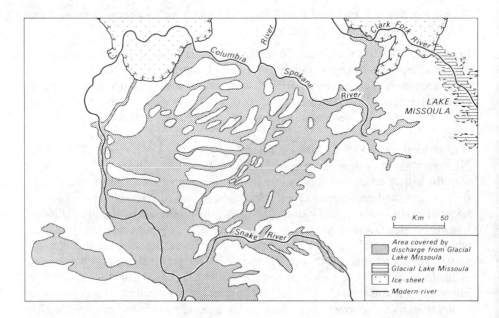

Figure 6.2 The Channelled Scablands of the Columbia Plateau, Washington
Source: after Baker 1978

Eddy Narrows on the Clark River, discharge rates must have been about 39.4 km³/ hour, with average velocity of about 70 km/hour. The maximum discharge of the modern Amazon would have been only about 1% of the discharge at Eddy Narrows. In view of the size of Lake Missoula, Pardee estimated that this discharge rate could not have been maintained for more than about one day.

The occurrence of cataclysmic events is frequently used as a basis for explanations in the earth sciences, even though the views of the catastrophists have apparently been refuted. It is as if the arguments between Buckland and Lyell and their followers had never occurred. Such examples appear to illustrate Feyerabend's view that no idea is so absurd that it can ever be discarded. Knowledge does not grow, Feyerabend suggests, by the development of a series of self-consistent theories which converge towards the 'truth'. Rather,

> . . . it is an ever increasing ocean of mutually incompatible (and perhaps incommen-
> surable) alternatives, each single theory, each fairy tale, each myth that is part of
> the collection forcing the others into a greater articulation and all of them con-
> tributing via this process of competition, to the development of our consciousness.
> Nothing is ever settled, no view can ever be omitted from a comprehensive
> account. (1975, p.30)

To suggest, as Popper does, that progress occurs in science by the process of conjecture and refutation is, Feyerabend claims, to take too simplistic a view. No idea can be rejected categorically. While the particular catastrophe of the Biblical Flood probably did not occur, rare large-scale geological events have been partly responsible for shaping the earth's surface. Not even all valleys can be explained by fluvial processes alone. In fact acceptance of the idea that cataclysmic events can sometimes be a significant factor in earth history has made geomorphologists look again at the uniformitarian principle and ask just what it is that is independent of time. The matter is considered in some detail by Simpson (1962), Kitts (1962), Rymer (1978), and Shea (1983).

Simpson (1962) concludes that the modern form of the uniformitarian principle is that we cannot suspend the operation of the physical laws which we assume govern the operation of geomorphological systems. At different places and at different times different processes may operate. But such differences are only due to changes in the initial conditions dictated by factors such as climate or geology. What is constant throughout time is the set of *laws* which govern processes. Shea (1983) rejects the notion of uniformitarianism altogether on grounds that it is a tautology, stating no more than that laws are universal statements.

These changes in the view taken of the uniformitarian principle could only have occurred in the context of ideas whch also involve the possibility of catastrophic events. The subversive attempts to use catastrophes to explain geomorphological phenomena, despite the refutation of the classical idea, has led to a more coherent formulation of uniformitarianism than would have been the case if the possibility of cataclysms had been forgotten. It is just this kind of experience that makes Feyerabend propose his first counterrule. Progress can only occur if a pluralistic methodology is pursued, and would be hampered if apparently refuted ideas were simply discarded.

The subversive nature of the process which Feyerabend envisages is well illustrated by the case of the Channelled Scablands described above. Writing in 1978, Bretz reflected on the reactions which his hypothesis of the Spokane Flood had brough forth:

> Catastrophism had virtually vanished from geological thinking when Hutton's concept of the 'Present is the key to the Past' was accepted and Uniformitarianism was born. Was not this debacle (i.e. the Spokane Flood) that had been deduced from the Channeled Scablands simply a return, a retreat to the dark ages of geology? It could not, it must not be tolerated.
>
> This, the writer of the 1923 article learned when, in 1927, he was invited to lecture on his findings and thinkings before the Geological Society of Washington, D.C. . . . A discussion followed the lecture and six elders spoke their prepared rebuttals. They demanded, in effect, a return to sanity and Uniformitarianism.
>
> The upstart theorist was not upset nor silenced . . . (p.1)

The controversy which followed is described in detail by Baker (1978) and Baker and Bunker (1985). Baker concludes:

> The Spokane Flood controversy is both a story of ironies and a marvellous exposition of scientific method. One cannot but be amazed at the spectacle of otherwise objective scientists twisting hypotheses to give a uniformitarian explanation for the Channeled Scablands. Undoubtably these men thought they were upholding the very framework of geology as it had been established by the writings of Hutton, Lyell, and Agassiz. The final irony may be that Bretz' critics never really appreciated the scientific implications of Agassiz' famous dictum 'study nature, not books'. (p.15)

Feyerabend's second counterrule, which is supposed to undermine our confidence in any scientific method, is that scientists should never be afraid to introduce theories which are inconsistent with established facts. Following the first counterrule, not only is pluralism desirable, it is inevitable because 'facts' cannot be used as the basis of theory choice.

The second counterrule arises because of the dependence of facts on theory. Feyerabend accepts this idea and uses it to argue that the clash between theory and fact is not always the fault of the theory, for facts are 'constituted' by older theories. Since those older theories may be wrong the clash does not matter.

According to Feyerabend, experience does not provide an unproblematic basis on which to make judgements between competing theories. No theory is ever in agreement with all facts. If refutation were the basis of theory choice then all theories ought to be rejected, and progress in science would be impossible. In fact, Feyerabend argues, scientists do not operate in this way. Wherever we look, Feyerabend suggests, we find examples of accepted theories clashing with facts, but the theory is not rejected. The anomaly is either ignored or explained away by ad hoc adjustments to the theory.

In support of his view Feyerabend (1975, 1978) cites the case of the Copernican revolution. At the time of Galileo, the Copernican view of the sun-centred solar system was so obviously at odds with accepted ideas that even Galileo thought it to be false. Galileo nevertheless pursued the matter until many of the contradictions were resolved. One may observe a somewhat similar pattern in the case of the development of fluvial theory by Hutton and his associates.

Although the idea of the Biblical Flood was rejected, the view that the landscape was the product of the slow and long-term operation of the fluvial processes was by no means compatible with all known facts. Once the ideas of the fluvialists were accepted, the drift deposits, once supposed to be evidence in favour of the Flood hypothesis, needed a separate explanation. This was no simple matter since glacial theory had not been developed. Lyell attempted to explain the origin of the drift by suggesting that they were rafted in by icebergs during a period of higher sea level (see Figure 6.3). Lyell believed that all Europe and North America, except the mountain tops, were once covered by the ocean and that during warmer phases swarms of icebergs would break away from the polar masses and move south. The debris-laden icebergs were beached on rocky islands and melting deposited the drift on the sea bed.

Part of the problem faced in explaining the drift in Britain was due to its often shelly character. This suggested to many that its origin must be connected with some marine agency. These early workers did not consider that the material was reworked marine deposits moved by ice sheets from the North Sea basin, but, as Chorley et al. (1964) note, Lyell was certainly well aware that moving ice could transport debris since he knew of Charpentier's observations in the Alps. They conclude: '. . . in many ways Lyell's theory was no more firmly founded on fact than was the theory of the diluvialists and at a later date Agassiz reserved equal criticism for them both' (p.202).

Lyell's adherence to the iceberg theory is even more intriguing when one considers that it involves appealing to periods of higher sea level while at the same time arguing against the occurrence of the Flood. The fluvialists did not object to marine inundations as such, but, it seems, merely to the suddenness of the Mosaic one. In fact, as Chorley et al. (1964 pp.165–169) show, having admitted the existence of periods of higher sea level, Lyell is led on to explain the origin of many features in terms of marine agencies, despite his strong fluvialist views.

What do such twists and turns in the history of geomorphological thought mean in terms of Feyerabend's attack on method? In many respects they seem to suggest that any rational reconstruction of the debate between the fluvial and diluvial schools would have to suggest that both theories ought to have been rejected, if the clash with facts is to be taken as the basis for theory evaluation. Yet, if refutation had been the guide to theory choice, then the progress which we see today would not have been

Figure 6.3 Some ideas on the origin of erratic boulders
                    Source: Prestwich 1886

possible. In fact such examples, Feyerabend might suggest, illustrate that the clash between theory and fact is really not very important at all. The conflict only occurs because the 'facts' depend on other equally speculative theories. There is no theory-neutral observation language that allows rational decisions to be made. There are merely alternative frameworks between which no real comparison is possible. The best that can be done is to develop these frameworks so as to remove any internal contradictions they may contain.

Feyerabend argues that our experience of the history of science shows that methodological rules of the kind suggested by Popper would eliminate science without replacing it with anything viable. The methodological rules are useless as an aid to science and, in any case, there has been progress without them. He adds:

In the past decade this has been realised by various thinkers, Kuhn and Lakatos amongst them. Kuhn's ideas are interesting but, alas, they are much too vague to give rise to anything but lots of hot air. If you don't believe me, look at the litera-ture. Never before has the philosophy of science been invaded by so many creeps and incompetents. Kuhn encourages people who have no idea why a stone falls to the ground to talk with assurance about scientific method. Now I have no objec-tion to incompetence but I do object to incompetence when it is accompanied by boredom and self-righteousness. And this is exactly what happens. We do not get interesting false ideas, we get boring ideas or words connected with no ideas at all. Secondly, whenever one tries to make Kuhn's ideas more definite one finds they are *false*. (1981, p.160)

Feyerabend regards Lakatos's ideas as altogether more sophisticated, yet they are no more than 'words which sound like the elements of a methodology'. In fact Lakatos provides no methodology at all. Feyerabend concludes – anything goes.

Against results

If science lacks any special method which distinguishes it from other activities, why are the products of science held in such great esteem by society? Like Zelinsky, Feyerabend claims that the view that science has a special place is a myth. It is merely the ideological pressures in society which make us listen to science to the exclusion of everything else.

... there does not exist a single argument that could be used to support the excep-tional role which science today plays in society. Science has done many things, but so have other ideologies. Science often proceeds systematically, but so do other ideologies... and besides there are no overriding rules which are adhered to in any circumstances; there is no 'scientific methodology' that can be used to separate science from the rest. (1981, p.162)

Following his critique of the limitations of science, Zelinsky (1974) concludes that the reason for the 'demigod's dilemma' is that in grappling with the complicated, abstracted, fluid and intersubjective data of the social sciences, man has been attempting to apply an inappropriate set of methods to the world of human beings. This view is illustrated most forcibly by a consideration of the problems faced by those who become involved in any programme of Environmental Impact Assessment (EIA).

Environmental Impact Assessment (Munn 1979) is a set of techniques which can be used to identify the environmental implications of major development projects such as a new dam, or a new airport, or some other large-scale engineering project. The techniques try to identify what environmental consequences will follow from any particular project, and to provide the kind of information needed by the planner in order to decide on the relative merits of different designs or locations and so on. EIAs are usually couched in objective, apparently scientific terms. Yet they face a fundamental difficulty. While they may be able to identify what change *will* follow from a particular project, they cannot answer the question of whether the project *should* go ahead. The decision to act is a subjective one. Questions of value cannot be resolved by scientific means. For many, such problems fatally undermine the usefulness of Environmental Impact Analysis (Chapman 1981).

In the face of the demigod's dilemma, Zelinsky (1974) suggests that, in those parts of geography involving the social world, we should treat with suspicion attempts to apply the scientific model. Equal insight can be achieved by other approaches, including intuitive ones. Feyerabend (1975, 1978) is more general in his conclusion. Not only is the 'rational scientific model' not applicable to social questions, *there is no such model even in the natural sciences*. He argues that science should be treated like any other myth which permeates society. If progress is to be achieved, he claims, there must be a formal separation between the state and science. Scientists may be consulted on important matters, but the final judgement must be left to democratically elected consulting bodies. In this way the views of science can be weighed against other views – and rejected if there seem good reasons for doing so. Other views of the world which conflict with science must be taught in schools and colleges and given equal emphasis. In short, society must not allow itself to be hijacked by the spurious claim of science to some superior kind of rationality. The extent to which such an attack on science can be maintained will be examined in the next chapter.

CHAPTER 7:
THE WORLD OF IDEAS

'. . . a comparison of the various frameworks is always possible'

Introduction

A major element in the debate about the methods of science concerns the need to resolve a simple but basic question: How can we justify any claim that we know something about the world? Many believe that science is special because it is better able than all other activities to justify its claim to know something. Scientists believe they can give good, rational reasons for holding one theory rather than another.

As we have seen, people have taken opposing views about the existence of a rational basis to science. On the one hand Kuhn and Feyerabend conclude that scientific judgement has no wholly rational basis. There is a pattern to the judgements which scientists make, and this pattern is shaped by accepted standards and values, but there is no rational justification for the standards and values themselves. Popper, on the other hand, not only disagrees, but identifies views such as those represented by Kuhn and Feyerabend as symptomatic of a fundamental philosophical problem:

> The main philosophical malady of our time is an intellectual and moral relativism, the latter being at least in part based on the former. By relativism . . . I mean here, the theory that the choice between competing theories is arbitrary; since either, there is no such thing as objective truth; or, if there is, no such thing as a theory which is true or at any rate (though perhaps not true) nearer to the truth than another theory; or, if there are two or more theories, no ways or means of deciding whether one of them is better than another. (1962, addendum)

Saying that the decisions of the relativist are arbitrary does not, of course, imply that choices between competing theories are made randomly, or according to some whim. If this were so then the claims of relativism would be absurd. For the relativist it is not that decisions are made without thought, but that they are made in the context of specific and particular social and intellectual frameworks. It is the claim that there are no grounds by which one framework can be justified rather than another that is the basis of relativism. In opposition to these ideas, rationalists such as Popper argue that there is such a thing as 'objective knowledge', and there is a rational basis for the choice between competing theories. This basis is provided by deductive logic and the ideas of falsification and falsifiability.

This chapter will present some conclusions from the debate that we have described in the last six. At the outset, it should be noted that there is no simple way to resolve the disputes we have considered. Philosophers have been, and continue to be, divided on many of these issues. All that can be done in a book such as this is to focus on those parts of the debate which are relevant for the everyday practice of physical geography, and to describe their implications.

Objective knowledge and world 3

Much of the appeal of Kuhn's ideas on paradigms and his views on the relationship of the scientist to some community-wide set of beliefs come from the recognition that scientists work in the context of a complex network of ideas. No problem can be considered in isolation. Other people may have examined it before. Most theories, even new ones, make assumptions which can only be justified on the basis of earlier work, or other people's experience, and so on. However, merely because the existence of background knowledge influences scientific activity, it does not follow that psychological and sociological forces are the most important ones shaping the growth of scientific knowledge. Kuhn is not the only philosopher who has considered the relationship of the scientist to the world of ideas. Popper provides an alternative explanation in terms of his 'three worlds'.

Without using the term world 'too seriously' Popper (1972b, p.106) distinguishes three distinct worlds: first, the world of physical objects (world 1); second, the world of consciousness and mental states (world 2); third, the world of 'objective contents of thought', or the world of ideas encoded in man's documents (world 3). He adds:

> Among the inmates of my 'third world' are, more especially, *theoretical systems*; but inmates just as important are *problems* and *problem situations*. And I will argue that the most important inmates of this world are *critical arguments*, and what may be called – in analogy to a physical state or to a state of consciousness – *the state of discussion* or the *state of critical argument*; and, of course, the contents of journals, books, and libraries. (p.107)

The world of physical states can interact with the world of consciousness. Through our senses we can form subjective impressions of the world around us. Similarly, the world of consciousness can interact with the world of ideas through the records left by other minds. Ideas once written down are there to be understood, misunderstood, believed, or discarded. Once put into a form that allows communication, however, ideas have an existence that is independent of the minds that created them.

In order to illustrate the 'autonomy' of world 3, and its independent influence on the life of man, Popper considers two thought experiments. One involves a situation in which all man's subjective learning was lost, but the contents of libraries and the capacity to learn from them was retained. The other involves the loss of both subjective learning *and* libraries and the capacity to learn from them. In the case of the first experiment civilization as we know it would eventually re-emerge. In the case of the second it would not.

The concept of a man-made yet autonomous world of ideas is an important element of Popper's philosophy of science, because it helps explain why the only rational attitude to adopt to scientific knowledge is a *critical* one. Consider again his simple characterisation of the scientific process (see Chapter 3):

$$P_1 \rightarrow TT \rightarrow EE \rightarrow P_2$$

where P_1 and P_2 are problems, TT, is some trial solution (or theory) and EE is the stage of error elimination. The process involves the interaction of each of Popper's three worlds. Often the initial problem arises out of the concerns of others. Similarly the trial solution may be one which others have used and encoded in world 3. Eventually, through the interaction with the physical world, the scientist can discover whether such ideas work for him, and his experience can be recorded in the world of ideas for others to consider.

Knowledge contained in world 3 is *objective* in the sense that it has an existence which is independent of the human mind. Popper makes no judgement about the truth or falsity of that knowledge. No knowledge is certain knowledge. World 3 may contain both insight *and* error. Each generation can modify or add to this body of knowledge on the basis of their experience of the world. However, some ideas, those which Popper attributes to science, have a particular property, which helps shape their development in a special way. This is the property of testability.

As we saw in Chapter 3, the critical rationalist argues that the property which distinguishes scientific statements from all others is that they are empirically testable, that is falsifiable. On the basis of the individual's own experience he can make a reasoned and public judgement about whether such a statement corresponds to the world as he observes it, and others can make judgements about such views. Although each human being is a prisoner of his own subjective consciousness, scientific ideas are *inter-subjectively* testable. Through *critical discussion* there can be some agreed and rational judgement formed about them.

What kind of relationship does the scientist enter into with the world of objective knowledge? What attitude and approach should be adopted towards it? We may answer such questions by considering once again the views of rationalism and the divisions between the rationalist and the relativist.

Views of rationality

In this book we have described two views of rationality: the classical and the critical rationalist. So far as the debate between them can be resolved, it can be decided fairly clearly in favour of the latter. The divisions between the two views are largely logical ones, and on logical grounds only the critical rationalists can escape major inconsistencies.

The principles of verifiability and induction are the two key ideas of the classical tradition. As we saw in Chapter 2, there is no logic of verification which would allow the truth of a statement to be established, in the sense that its truth is beyond doubt. Similarly, there is no way in which observations, however true, can be assembled to justify any theory or law. If there is anything which is certain in the methodological debate described above it is that the classical approach cannot provide any rational justification for holding one theory rather than another.

In contrast to the classical tradition, the logical framework provided by critical rationalism is more secure. Few dispute the logic embodied in the principle of falsification. Although no number of observations can justify the truth of a generalization, only one counter-observation is required to refute it. Using such an idea one may exploit the power of deductive logic to test a theory.

More importantly, the deductive method provides the logical framework within which theories can be pitted against each other. As Chapter 3 shows, a theory can be tested by deciding what consequences must follow if it is true. If the results of an experiment or a set of observations run counter to what is predicted on the basis of the theory under test, then the scientist must make a judgement about what set of ideas is most reasonable. Either the theory is wrong, or the observations and experiments are faulty. *Rationality demands that both cannot be true.* The idea which better fits the facts is retained. It is on the basis of such an approach that the scientist must justify why he holds one theory rather than another. In a world where no knowledge is certain, an approach to science based on the careful and systematic criticism of ideas provides a rational basis for action:

> It may now be possible for us to answer the question: How and why do we accept one theory in preference to others?
>
> The preference is certainly not due to anything like an experiential justification of the statements composing the theory; it is not due to a logical reduction of theory to experience. We choose the theory which best holds its own in competition with other theories; the one which, by natural selection, proves itself the fittest to survive. This will be the one which not only has stood up to the severest tests, but the one which is also testable in the most rigorous way. A theory is a tool which we test by applying it, and which we judge as to its fitness by the results of its applications. (Popper 1972a, p.108)

Rationalism vs relativism

Having established just what the rationalist image of science involves, we may turn to the second major theme examined in the previous six chapters, namely the extent to which rationalism can provide a complete account of scientific method. Relativists, such as Kuhn and Feyerabend, do not dispute the logical framework which Popper proposes, but merely its relevance. They start from the observation that rarely do theories agree with all known facts. Thus on logical grounds most theories should be refuted. They suggest that history shows that scientists are much more conservative in their attitude. Theories are retained despite known anomalies. Thus they claim that there must be forces other than logic which shape science. According to Kuhn and Feyerabend, ideas are pursued, discussed and promoted for social, psychological and political reasons, and judgements are made about such ideas according to these intellectual frameworks.

In attempting to resolve the debate between rationalism and relativism it is important to note that Popper does not deny that there are sociological and psychological dimensions to scientific judgement. According to the critical rationalists, the judgement which the scientific community makes about a theory depends on the view it takes of various empirical tests and this involves factors other than logic. However, merely because decisions are affected by criteria other than logical ones, it does not follow that science lacks a rational basis.

Popper is a realist in the sense that he believes that things described by theories actually exist. He also maintains the correspondence theory of truth. A statement is either true or false according to whether it matches the way things actually are. Although the scientist can never know for certain whether theories are true, judgements can be made about the match on the basis of observation and experiment. The judgement involves decisions about matters of fact and arguments about what consequences must follow from such decisions.

In order to illustrate the way in which the scientific community makes a judgement about the validity of a theory, Popper uses the analogy of trial by jury:

> The *verdict* of the jury . . . like that of an experimenter, is an answer to a question of fact . . . By its decision, the jury accepts by agreement, a statement about a factual occurrence – a basic statement, as it were. The significance of the decision lies in the fact that from it, together with the universal statements of the system (of criminal law) certain consequences can be deduced . . . the verdict plays the part of a 'true statement of fact'. But it is clear that the statement need not be true merely because the jury has accepted it. This fact is acknowledged in the rule allowing a verdict to be quashed or revised. (1972a, pp.109–110)

The verdict reached by the jury may reflect subjective conviction, even personal bias. But it can be challenged, because subjective factors do not *justify* the decision reached. The verdict is governed by the effort to discover the objective truth. Thus it can be questioned on the basis of whether it was reached using procedures which are likely to discover the truth. In a similar way the view which the scientific community takes of a particular experimental test can be challenged if it does not conform to such rules of procedure.

Popper pursues the analogy between the evaluation of a theory and trial by jury in order to show how, once taken, the decision about the matter of fact then forms the basis of judgement as a whole:

In contrast to the verdict of the jury, the *judgement* of the judge is 'reasoned'; it needs, and contains, a justification. The judge tries to justify it by, or deduce it logically from, other statements: the statements of the legal system, combined with the verdict that plays the role of initial conditions. (1972a, p.110)

Thus, if the law specifies a particular punishment for a given crime, and if the verdict is 'guilty', certain consequences follow. Similarly, once a particular view is formed about an experimental test of a theory, then there are certain logical consequences for the scientist: the theory is either corroborated or refuted.

We may retain the analogy in order to bring out the difference between Popper's rationalism and the arguments of the relativists. According to Kuhn and Feyerabend, judgements can be made about theories, but the criteria applied to decide matters of fact have no rational basis. They simply depend on the prevailing community standards at the time. It is as if each jury in a trial makes its decision about statements of fact on its own criteria, and these decisions cannot be compared and none have any prior claim over the others. In opposition Popper holds that there is a basis for comparison.

As we have seen in Chapters 4 and 6, both Kuhn and Feyerabend base much of their argument on the idea of incommensurability. Each new theory provides a new framework with which to interpret the world and there is no neutral language in which to express and compare competing theories. It is necessary to ask, therefore, to what extent is incommensurability a serious problem for the rationalist.

In Chapter 4 we distinguished three ways in which incommensurability could arise, these were: topic-incommensurability, dissociation and meaning incommensurability. Only the last would seem to pose any potential threat for the rationalist.

Topic-incommensurability simply means that different theories or even different research programmes cannot be compared because they tackle different problems. As in the case of the debate between the proponents of the process-response approach and historical geomorphology (see Case Study 4.2), there is no factual basis on which to make a comparison between them because they pursue different ends. Preference is guided by the fact not that one solves problems that the other cannot, but that one leads to questions that seem more interesting or fruitful than the other. That scientists can choose, subjectively, to follow one fashion rather than another is no problem for the rationalist. The origin of problems, like the origin of theories, is not a concern of methodology. An analysis of scientific method involves only the way in which questions are answered and whether one solution stands up better than another. Thus the only way in which topic-incommensurability would be important would be if it could be argued that with a change in outlook, standards used in rational argument also changed. Such a thesis is difficult to support.

Nor does incommensurability in the form of dissociation pose any serious problem for the rationalist. Conceptual frameworks may change so that older views appear incomprehensible at some later stage, but there is nothing in principle to prevent a later generation of scientists trying to reconstruct the 'situational logic' of earlier workers by an analysis of any relevant literature which remains. This is a standard task for everyone who investigates the contents of world 3. In some situations all the relevant material may not be available, and the reconstruction cannot be made. This is not a problem of logic, but of history.

Only meaning-incommensurability appears to pose serious problems for the rationalist. The idea of meaning-incommensurability is that the meaning of terms is so closely connected with theory that comparison between theories cannot be made because there is no common language in which they can be compared. Even if theories appear to tackle the same problem they cannot be compared because the language they use to describe reality differs in subtle ways.

There are various reactions to the problem of meaning-incommensurability. Some have suggested that the whole idea is incoherent and misleading. How can one speak of competing theories if we do not accept that they are talking about the same thing, and are therefore comparable?

> Taken literally, it is implausible because it suggests that I could never have any justifiable grounds for holding any belief whatsoever, say, that I now see a type-writer, rather than a belief incompatible with it . . . if I could have grounds for preferring one of those beliefs to the other, why could I not have grounds for preferring one theory to another? (Newton-Smith 1981, p.148)

Chalmers (1982, pp.137–138) suggests that even if theories are incommensurable it does not follow that they cannot be compared. The scientist could confront each theory with a set of tests defined in its own terms and evaluate each theory in isolation. A judgement could then be made on the basis of which theory stood up best to critical evaluation. Newton-Smith (1981) points out that if, in fact, theories are incommensurable, why even try to choose between them? Why not believe them all?

The problem of incommensurability is complex but it is not one that need detain us here (a fuller discussion can be found in Hacking 1983). In the end it seems that the attitude one adopts towards the problem depends on the view one takes of the thesis of realism. If, according to realism, there is a world of things independent of men's minds, and if people can refer to those things, then, although the terms of those references may change as ideas change, the different images or stereotypes of the world can be compared.

Consider, for example, the idea of a 'land bridge' in the context of the stabilist and mobilist theories of the earth (see Case Study 4.1). Here we have a term which changes its meaning dramatically depending on which theory is adopted. According to the stabilist view, a land bridge is an independent landmass which has subsided to sever the connection between continents. According to the mobilist view it is a connection formed by the juxtaposition of existing continental masses. Although the meaning of the term land bridge changes between theories, it does not follow that the ideas are incommensurable. Each stereotype has certain consequences which can be tested. Each, for example, implies something different about the structure of ocean basins.

In attempting to evaluate the claims of the relativists it is clear that the idea of incommensurability simply does not fit in with the belief that there is a reality to be investigated:

> A critical discussion and a comparison of the various frameworks is always poss-
> ible. It is just dogma – a dangerous dogma – that the different frameworks are like
> mutually untranslatable languages. The fact is that even totally different
> languages (like English and Hopi, or Chinese) are not untranslatable, and that
> there are many Hopis or Chinese who have learnt to master English very well . . .
> A new insight may strike us like a flash of lightning. But this does not mean that we
> cannot evaluate, critically and rationally, our former views in the light of new
> ones. (Popper 1970, pp.56–57).

If one can accept that a comparison between conceptual frameworks can be achieved through our experience of the world, then there seems little to support the ideas of relativism.

The lessons of history

Although it may be argued that the problem of incommensurability is not a serious one for the rationalist, problems of the fit of the rationalist accounts to the history of science remain. We will examine two aspects. The first involves the climate in which new ideas are introduced into science. The second considers the conservative behaviour which scientists sometimes exhibit.

1. The myth of refutation

Feyerabend (1975, 1978) claims that the idea of scientific method involving firm, unchanging and absolutely binding principles for conducting the business of science meets considerable difficulty when confronted with the results of historical research. He bases much of his argument against method on the obervation that the history of science contains many examples of scientists who have held views, or pursued theories, which were obviously falsified or which were counter to accepted beliefs, and that this, nevertheless, led to success. Are such arguments sufficient to under-mine the idea that science has a rational basis?

According to Feyerabend (see Chapter 6) rationalist methodologies argue that judgements about new theories are made on the basis of their consistency with currently accepted theories and results. However, Feyerabend argues that there is no rule which we can find which has not been violated at some time. 'Counterinduction', involving the scientist behaving in the opposite way, is equally likely to lead to progress.

Consider the two counterrules suggested by Feyerabend, namely that the scientist should introduce theories which are (a) at odds with currently accepted theories and (b) at odds with currently accepted results. Are these rules so outrageous that they shake our confidence in a rational basis for science? In fact, neither rule describes particularly irrational behaviour. Indeed, the critical rationalist would encourage such attitudes, since it is only through the comparison with other, competing ideas, that any theory can be tested. Theories should be consistent, but the scientist should not promote a single world view. What becomes the 'conventional wisdom' of the time ought not to arise through a lack of imagination, but through the critical evaluation of ideas.

The fact that scientific progress does not depend on rational behaviour at every step does not mean that the whole enterprise of science is irrational. We may agree with Feyerabend that there is no idea, however ancient or absurd, that may not be revived and eventually lead to new knowledge. The scientist is free to examine the contents of the world of objective knowledge (world 3) and, if possible, test out these visions of reality. Consider the case of plate tectonics. As we have seen (Case Study 4.1), this theory grew out of Wegener's hypothesis of continental drift and other even earlier ideas. Yet the fact that earlier workers held these ideas, despite their obvious conflict with accepted views, does not mean that the progress we now perceive was somehow irrational. Sponsorship in the 1960s may have provided the stimulus for acceptance of the 'mobilist paradigm', but the argument of Hallam (1973) and Jones (1974) that the idea was accepted more for sociological reasons than for logical ones is absurd. The choice between the competing stabilist and mobilist programmes was guided by an examination of their logical consequences. A similar point could be made using the case of the Spokane Flood controversy (see Baker 1978, and Chapter 6, p.106–108; see also Exercise 7.1).

It is often suggested that ideas which are apparently mistaken or which appear to lie outside science have little value. The logical positivists, for example, suggested that only those statements which are empirically verifiable have meaning. Although the critical rationalist also has a criterion which distinguishes scientific statements from all others (the principle of falsifiability), it does not have the implication that ideas which are outside science have no value. Indeed Popper views these 'metaphysical ideas' as being extremely important for science:

> ... it cannot be denied that along with metaphysical ideas which have obstructed the advance of science there have been others ... which have aided it. And looking at the matter from the philosophical angle, I am inclined to think that scientific discovery is impossible without faith in ideas which are of a purely speculative kind, and sometimes even quite hazy; a faith which is completely unwarranted from the point of view of science ... (1972a, p.38)

The fact that the scientist may choose to work with ideas that are apparently at odds with accepted theories and results does not mean that science has an irrational basis. The scientist must be prepared to criticize his theories, and though they may be refuted, there is no reason to discard them too casually. Most theories need to be developed. There is nothing to guarantee which ideas will eventually be successful, but only through criticism can progress be made. It is a myth that refuted ideas have no value and must be discarded. The scientist is continually forced to work with ideas which are apparently refuted by observations. It is out of such experience that progress develops.

2. Normal science vs critical science

A second line of attack on the idea that science has a rational basis comes from Kuhn's observation that most scientists do not usually question their conceptual framework. As normal scientists (see Chapter 4), they behave more conservatively. They are content merely to solve puzzles posed by the paradigm rather than question basic assumptions. Does the existence of normal science undermine the idea that science has a rational basis?

In an analysis of what he sees as the dangers of normal science, Popper (1970, 1983) acknowledges that normal science does in fact exist. There are, he suggests, many examples of people who accept ideas uncritically and never challenge the conceptual framework in which they conduct their research. Although scientists may have behaved in this way, and may continue to do so, he argues that their activities constitute no more than *bad science*. The fact that scientists may accept ideas uncritically is not sufficient to undermine the idea that science has a rational method. People can and perhaps always will make decisions for irrational reasons. However, merely because it is possible to act irrationally, it does not follow that we should choose to be irrational.

These arguments can be extended to Lakatos's idea of a research programme with its notion of a set of irrefutable basic assumptions represented by the 'hard core' (see Chapter 5). In any programme of research it is certain that there will be facts or theories which go unquestioned. Yet the suggestion that there *ought* to be such ideas is extremely dangerous. For, once ideas are institutionalized, they are never criticised. It may make for easier research, but not research that is likely to yield many new insights. Lakatos acknowledges that the method of critical testing is supposed to operate in developing the auxiliary hypotheses which protect the hard core. Why should certain ideas be insulated from the process? In the end, the choice between alternative research programmes has to be based on criticism of the hard core, along the lines suggested by Popper. It seems unnecessary to postpone the evaluation by some kind of 'methodological decision' made by the scientific community.

The potential dangers of the attitude implied by the idea of normal science may be illustrated by Brunsden and Thornes's (1979) account of some of the basic assumptions of modern geomorphology. Brunsden and Thornes identify four basic principles (see Case Study 5.2). These may represent the basic assumptions of the process-response approach, but hardly seem an adequate basis on which to conduct 'realist science', since as we saw in Chapter 5 they are simply not falsifiable. If no observations could be made which could show any of the propositions to be false, on what rational grounds are these ideas to be judged? Is it rational to base geomorphological realism on such untestable ideas? Will later generations of geomorphologists write, as Bishop (1980) has of Davis's cycle, that such basic principles were accepted uncritically? Rather than being content to suggest that the principles 'may or may not' apply, a major contribution might be made if it could be shown under what circumstances they do not apply. At least then we would have some testable ideas about the constraints which operate in geomorphological systems.

The role of methodology

The responses to the arguments of the relativists outlined above bring out the differences in approach of the various schools of thought. Kuhn, Feyerabend, and to some extent Lakatos, base their accounts of science on the lessons which history teaches us, and the implication is that future generations of scientists should strive to behave in similar ways. Popper, on the other hand, does not base his arguments on history alone, but considers methodology in terms of logic (1983, p.xxv). An account of the history of science, he would suggest, is no basis on which to prescribe how the scientist should act.

Burke described the dichotomy between the descriptive and prescriptive approaches as follows:

> ... Popper's philosophy of science ... is determined not primarily by what, as a matter of history or sociology, scientists actually do or profess to do, but rather by an analysis of what, as a matter of logic, they can and cannot do. (1983, p.103)

And it is in this light that it should be judged.

Moss (1972) has argued that, in geography, the study of methodology ought to play a clarifying role, but that it has no prescriptive or normative functions. Although elsewhere (Moss 1977, 1979) he has argued for the deductive method, he appears not to appreciate the implications of the philosophical position which he has adopted. An appreciation of methodology can only play a clarifying role if its prescriptive elements are emphasized.

The methodology described by the critical rationalist is both prescriptive, saying how the scientist ought to proceed, and restrictive, suggesting what approaches are untenable. Thus, on the one hand it emphasizes that the scientist should proceed by the method of conjecture and refutation, that is, critically evaluating the deductive consequences of his theory. On the other hand, it argues that justification founded on the notions of 'verification' or 'repeated observations' is mistaken. Although it points out what, as a matter of logic, the scientist can and cannot do, it should not be looked on as a set of rigid and binding rules. Popper writes:

I assert that no scientific method exists ... To put it in a more direct way:

(1) There is no method of discovering a scientific theory.
(2) There is no method of ascertaining the truth of a scientific hypothesis, i.e. no method of verification.
(3) There is no method of ascertaining whether a hypothesis is 'probable', or probably true.

But he adds:

I am a rationalist. By a rationalist I mean a man who wishes to understand the world, and to learn by arguing with others ... By 'arguing with others' I mean, more especially, criticizing them; inviting their criticism; and trying to learn from it ... I believe that the *so-called method of science consists in this kind of criticism.* Scientific theories are distinguished from myths merely in being criticizable, and being open to modification in the light of criticism. (1983, pp.6–7)

The methodology represented by the ideas of critical rationalism represents an attempt to formalize the bases on which rational judgement can be made. It is not a recipe for the production of scientific knowledge that can be employed with the minimum of thought. It describes no more than a framework in which knowledge is more likely to grow.

The evaluation of ideas in science involves the scientist in matters of judgement as well as logic. Although scientific theories must be testable, it is naive to assume, for example, that a theory is discarded as soon as any counter-evidence emerges. On logical grounds, one single counter-observation is sufficient to falsify a scientific law. However, the conclusion that the law or theory is in fact falsified requires the scientist to make judgements that cannot be decided by logic alone. The judgement depends on the view which the scientist takes of the law or theory and the apparently refuting evidence. This view may be shaped by other tests and observations. There is no method by which some observations can be made more secure than others. In the end, it is only on the basis of experience that a judgement can be made about the statements of fact obtained from the refuting experiment. But once that judgement is made, the logical principles of the deductive method specify what conclusions must follow. Either the experimental evidence is accepted and the theory refuted, or the observations are discarded.

The fact that there is no way in which judgements about questions of fact can ultimately be verified means that there is nothing that is certain in science. Although objective knowledge is not certain knowledge the scientist can, nevertheless, justify decisions about theories on rational grounds, since some ideas have withstood criticism better than others.

In reviewing the claims and counter-claims of the last six chapters it would seem that a strong case can be made for the belief that science has a rational basis, and that this basis is provided by something like Popper's critical approach. Certainly it seems a more practical basis for action than any of the other positions. To claim that in science 'anything goes' is to say very little that is helpful in finding out about the world.

The view that science has a rational basis is probably one which would be accepted by most physical geographers. Yet although they have increasingly come to view their work in terms of 'realism' and 'the hypothetico-deductive method' very few seem aware of the implications which follow from these ideas. In the first part of this book we have described the character of the rationalist approach to science. For the practising scientist none of this would have any value unless the ideas can be translated to everyday work. In the second part of this book we will describe the implications which follow from adopting the critical rationalist approach for activities such as theorizing, modelling, classifying, measuring, and experimenting.

PART 2:
PRACTICE

CHAPTER 8:
THEORIZING

'. . . a series originating in and repeated to infinity'

Introduction
The first part of this book was concerned with the different views which have been taken of science. Our conclusion was that the approach based on deductive reasoning was perhaps the most acceptable. The second part of the book examines the practical implications of these ideas for the way scientists go about observing the world and testing theories. Behind all the approaches to science which have been analysed in the first part of this book is the assumption that it is theory which provides the framework for understanding the world. The creation of theories, however, is not subject to the same kind of rational analysis as their evaluation. In order to explore the process of theorizing, this chapter focuses on two issues, the origin of theories and their development.

The origin of theories
It has proved difficult to describe how a scientist comes up with a theory. As we have seen in Chapter 3, for example, Popper regards questions about the origin of theories as questions of psychology. Theories may have their origin in logical argument, inspiration, bias, the misunderstanding of other people's ideas or a general antithesis to them, so that it appears as though there is little that is constructive to be said. In fact, the impression that one gains from the various accounts of science is that, in the context of theory development at least, the maxim 'anything goes' is a good one. However, none of this helps the student who has to make sense of the contemporary situation and rationalize the origin and development of theories. Although it has to be admitted that theorizing falls outside the realm of logic, it may be useful to regard it as a craft of the scientist which can be learned and improved (Platt 1964). In order to do this it is necessary to understand the context in which theorizing takes place.

Problem situations
Theories are developed to explain what we observe. The question which can be asked therefore is: How do we jump from an observation statement to a theory? This ques-

tion is misleading since it begs an inductive answer. In response we should say that
the jump is not from an observation statement but from a *problem situation* provided
by the observations. Any theory subsequently developed must allow us to explain the
observations which created the problem.

In any problem situation a number of theories both good and bad can be invented
to explain observations. The problem then becomes: How do we jump to a good
theory? The answer is by jumping first to *any* theory and then testing it to find out
whether it is good or not (Popper 1972a, p. 55). The testing of a theory will then raise
new and unexpected problems, like the one which gave rise to the situation in the first
place. It is this sequence of development which best characterizes science as '*progressing from problems to problems* – to problems of ever increasing depth' (ibid., p.
222). Science starts with problems which may arise from contradictions of a theory or
from some previously held expectation. The problem makes us conscious of our
theories and challenges us to learn by observing and experimenting. Every
worthwhile new theory creates new problems and it is through these problems that it
is fruitful.

This progression is represented by Popper's schema described in Chapters 3
and 7:

$$P_1 \rightarrow TT \rightarrow EE \rightarrow P_2$$

where P_1 and P_2 are problems, TT is a tentative theory and EE is the stage of error
elimination. This schema represents a continuing chain of events. Any new theory
cannot be understood except within the context of such a chain. As James Joyce
expressed it, although about matters very different:

> ... each one who enters imagines himself to be the first to enter whereas he is
> always the last term of a preceding series even if the first term of a succeeding one
> each imagining himself to be the first, last only and alone, whereas he is neither
> first nor last nor only nor alone in a series originating in and repeated to infinity.
> (1960, p. 652)

The implication of Popper's schema is that in coming to appreciate the origin of
theories we must appreciate their ancestry. The advice 'to take any theory' is used to
illustrate a point about scientific method, that the key factor is the elimination of
error through criticism. It is hardly realistic advice for theory development because,
as it is stated, it ignores the problem situation and the antecedence of ideas. It is open
to anyone, however, in a problem situation to propose a theory which has a pedigree
in a different tradition from the theory which gave rise to the problem in the first
place. Indeed such new theories can provide entirely new ways of looking at problems
and bring to bear the accumulated experience of problem situations from elsewhere.
Gilbert (1896), in trying to understand the origin of hypotheses, suggested that the
scientist who was most widely read and experienced in solving the problems of
several disciplines would be the best generator of hypotheses. He believed that this
was because hypotheses were generated by analogy. Having the widest experience of

abstract ideas promoted a greater capability to generate them in new problem situations.

In order to understand further the problem of the origin of theories examine the problem situations described and referred to in Case Study 8.1. It will be discussed in the section which follows.

Case Study 8.1: Problem situations – ecosystem development

The ideas of succession and ecosystem are key concepts in ecology. They have been discussed by ecologists throughout the twentieth century. In this case study we will consider one or two steps in the recent development of these concepts. Previous steps are reviewed in Colinvaux's (1973) excellent review of ecological ideas.

In the late 1960s Odum (1969) sought to summarize those ideas about succession and ecosystem development in terms of systems theory. Odum defined succession as an orderly process of community development which resulted from the modification of the physical environment by the community. Succession resulted in a stabilized ecosystem with maximum biomass and maximum symbiotic interaction between organisms. He argued that the whole process was reasonably predictable so that in similar situations similar successions will occur. In all he put forward 24 specific hypotheses about the different aspects of ecosystem development during succession (see Table 8.1 and Figure 8.1). In relation to the way in which ecosystems varied in their pattern of nutrient losses Odum concluded:

> An important trend in successional development is a closing or 'tightening' of the biogeochemical cycling of major nutrients such as nitrogen, phosphorus and calcium. Mature systems as compared to developing ones, have a greater capacity to entrap and hold nutrients for cycling within the system. (pp. 266–267)

Vitousek and Reiners (1975) were critical of Odum's conclusion about nutrient cycling. They pointed out that it was inconsistent with another of his hypotheses which stated that in a climax ecosystem net production is eventually zero. They argue:

> ... an ecosystem will show an excess of inputs for a particular element over outputs roughly in proportion to the rate at which that particular element is bound into net production ... After peak net increment has passed this difference will decline towards zero. (p. 376)

Thus Odum's idea that mature ecosystems have the greatest ability to entrap nutrients because of their large biomass appears to conflict with the idea that once net productivity falls to zero, nutrient outputs will rise to match inputs. Vitousek and Reiners's hypothesis is summarized in Figure 8.2. In support of their hypothesis they present data on the chemistry of streams draining developing and mature forests and

Table 8.1 Trends to be expected in the development of ecosystems.

Ecosystem attributes	Developmental stages	Mature stages
Community energetics		
1. Gross production/community respiration (P/R ratio)	Greater or less than 1	Approaches 1
2. Gross production/standing crop biomass (P/B ratio)	High	Low
3. Biomass supported/unit energy flow (B/E ratio)	Low	High
4. Net community production (yield)	High	Low
5. Food chains	Linear, predominantly grazing	Weblike, predominantly detritus
Community structure		
6. Total organic matter	Small	Large
7. Inorganic nutrients	Extrabiotic	Intrabiotic
8. Species diversity – variety component	Low	High
9. Species diversity – equitability component	Low	High
10. Biochemical diversity	Low	High
11. Stratification and spatial heterogeneity (pattern diversity)	Poorly organized	Well organized
Life history		
12. Niche specialization	Broad	Narrow
13. Size of organism	Small	Large
14. Life cycles	Short, simple	Long, complex
Nutrient cycling		
15. Mineral cycles	Open	Closed
16. Nutrient exchange rate, between organisms and environment	Rapid	Slow
17. Role of detritus in nutrient regeneration	Unimportant	Important
Selection pressure		
18. Growth form	For rapid growth ('r-selection')	For feedback control ('k-selection')
19. Production	Quantity	Quality
Overall homeostasis		
20. Internal symbiosis	Undeveloped	Developed
21. Nutrient conservation	Poor	Good
22. Stability (resistance to external perterbations)	Poor	Good
23. Entropy	High	Low
24. Information	Low	High

Source: after Odum 1969

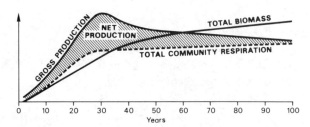

Figure 8.1 Energetics of succession in a forest
Source: after Odum 1969

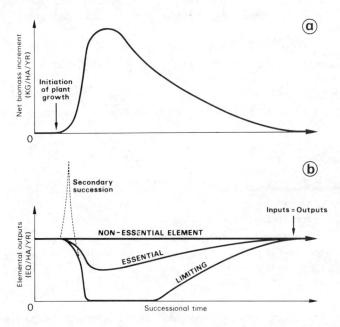

Figure 8.2 Variations of biomass increment and nutrient loss with time in successional ecosystems
Source: after Vitousek and Reiners 1975

show that the nutrient loss from the mature system is higher. However, they also report data from the literature which contradicts their hypothesis:

> A recent non-comparative study of nitrogen losses from a 450 year old Douglas Fir dominated watershed would seem to be a more severe test of our hypothesis. Fredrikson (1972) reported that losses were low for what might seem to be a steady state system. (p. 380)

Their reaction to the contradictory evidence is not simply to reject their hypothesis. Instead, they consider how the results could be obtained without rejecting it by calling into question the possibility of finding uniform stands of mature forest, and their strategy is to redefine their system of interest as being a steady state system and not a mature forest:

> ... there should be no confusion between mature or climax in the phytosociological sense and steady state in the ecosystem sense ... statements to the effect that mature ecosystems are 'tight' whereas disturbance can lead to 'leakage' are misleading ... steady state by definition must include elemental loss ... (pp. 380–381)

Finally they restate their adherence to their original hypothesis:

> We maintain that intermediate aged successional ecosystems will have lower nutrient losses than either very young or very old (mature) ecosystems. (p. 381)

Consider the material of this case study.

Summarize the successive problems faced in turn by Odum and Vitousek and Reiners.

Discuss the place of systems thinking in the development of Odum's ideas and Vitousek and Reiners's reaction to them.

Discuss the overall status of Odum's model in the light of Vitousek and Reiners's findings.

From the work of Odum and Vitousek and Reiners it is possible to trace something of the problem situation in which their ideas developed. Odum's original paper was prompted by the general concern that man would eventually outstrip the available resources of the biosphere. Such concerns stimulated him to attempt an overview of ecosystem development so that more effective management techniques might develop. Odum's problem situation was also determined by the historical development of ideas which made up successional theory (Colinvaux 1973). He considered that they needed to be brought together and summarized more precisely. Systems theory provided him with a set of analogies to structure his view of vegetation communities. The climax condition is seen as analogous to all systems at equilibrium.

Vitousek and Reiners's problem situation arose from what they saw as a logical inconsistency in Odum's ideas. Although they do not acknowledge the use of analogy it is employed through systems thinking. The inconsistency was recognized through a consideration of system nutrient budgets. Input and output must balance for any system in equilibrium. Although Odum had used systems analysis he did not extend it to consider nutrient budgets which would have exposed his error. This inconsistency which Vitousek and Reiners recognized led them to consider Odum's hypothesis in relation to field data. It is interesting to note that their work did not result in a simple corroboration or refutation of Odum's ideas but in an attempt to reformulate basic concepts of communities. They found evidence to refute both their own and Odum's hypotheses. As a result they call into question the possibility of finding uniform stands of mature forest. In applying systems thinking to the problem of climax communities they were led to reformulate their concepts of the way in which vegetation behaves. Given such ideas, later workers face a problem situation radically different from that which faced Odum or Vitousek and Reiners. Their problem is to test whether the patchiness of mature stands of vegetation has an effect on overall nutrient output and whether or not a mature community is ever in a steady state.

Work such as that described in Case Study 8.1 illustrates that theory development does not occur spontaneously. Theories develop within a context.

All growth of knowledge consists in the improvement of existing knowledge which is changed in the hope of approaching nearer the truth. (Popper 1972a, p. 71, italics original)

The growth of knowledge arises out of problem situations which arise from trying to explain the world. The modifications made to ideas can be additions to or negations of previous knowledge whether it be theories or metaphysical ideas or inborn dispositions. In coming to understand the recognition of problem situations, the replacement of theories and the origin of theories, we must examine the relation between them and other members of world 3.

Our analysis of the idea of problem situations illustrates that a theory arises in the context of a complex framework of ideas which includes other scientific theories. Partly, however, it occurs in response to ideas of a more tenuous kind. Such ideas are often referred to as metaphysical. We will refer to them as *myths*. These two influences on theory development are considered in the sections which follow.

Theories and meta-theories

In testing theories there are wider bodies of theory which are generally not questioned. Indeed these wider bodies of theory are used as though they are true (see Chapters 4, 5 and 7). Thus a geologist may examine a particular theory about transcurrent faulting on ocean floors, recognizing that this theory is contained within the wider theory of plate tectonics. A biogeographer may devise a theory to explain the distribution of a particular organism, and this theory is embedded in the broader theory of evolution. When theories are nested in this way we can describe the larger theoretical frameworks as *meta-theories*. Meta-theories are part of the background knowledge which any scientist brings to bear on a problem and which is contained within world 3. We can illustrate the relation between theories and meta-theories by the material of Case Study 8.1 in which Odum's and Vitousek and Reiners's theories are embedded in a wider framework of ecosystem and successional theory. Meta-theories are not myths because they are falsifiable. What concerns us here is the status of meta-theories in the context of working theories being falsified, since this exposes the relations between the two.

In Case Study 8.1, if it were found that nutrient losses from mature stand eco-systems were universally less than from medium aged stands, what would this imply for Odum's more general ideas about succession? The answer to this question is 'it depends'. It depends on whether the test of the working theory had been devised in such a way as to have implications for the meta-theory or not. In the case of Vitousek and Reiners's hypothesis it is clear that its failure would have considerable impli-cations for successional theory, since the only way in which they explain their results is to say that succession has not occurred. If Vitousek and Reiners had explained their results by some mechanism which was consistent with ecosystem development, then the larger theory might not be in danger. Even if a meta-theory is not refuted every failure of a working theory must have implications for the meta-theory within which it is nested. The problem is in judging where the problem lies.

Following on from Case Study 8.1, if it is found that forest succession does not fulfil the predictions that the specific theory of succession in such sites predicts, then this has implication for the whole of successional theory. Yet reformulation of the meta-theory would not necessarily take place simply as a response to an isolated refutation. We have considered the nature of such judgements in Chapter 3 under 'Falsification and falsifiability'. There are many possible reasons why a theory may not be seen to fit observations and a possible fault with the underlying theoretical framework is but one. The judgement of scientists is to analyse the problem situation and to decide where the solution to it may be. Thus in seeking to understand the origin of theories the possibility must always be considered that the problem situ-ation derives from the failure of a meta-theory.

Myths and the conceptual base

Myths are another part of the framework of ideas out of which theories develop. In order to appreciate the distinction between theories and non-testable myths and their respective roles in physical geography consider the extracts in Case Study 8.2. Each has been held to be an important consideration to theory in physical geography (see, for example, Tinkler 1985). You might also try Exercise 8.1.

Case Study 8.2: Myths in geomorphology

Consider the following three extracts and assess the extent to which the ideas expressed in each of them are potentially falsifiable. Try to assess the significance for modern geomorphology of the work from which each extract is taken. You might like to try Exercise 8.1.

Extract 1 Time, space and causality

The distinction of cause and effect among geomorphic variables varies with the size of a landscape and with time. Landscapes can be considered either as a whole or in terms of their components, or they can be considered either as a result of past events or as a result of modern erosive agents. Depending on one's viewpoint the landscape is one stage in a cycle of erosion or a feature in dynamic equilibrium with the forces operative. These views are not mutually exclusive. It is just that the more specific we become the shorter is the time span with which we deal and the smaller is the space we can consider.

The time span considered also influences causality, as the sets of independent and dependent variables... show [Table 8.2]. If the variables were not considered with respect to the time span involved, in many cases it would be difficult to determine which variables are independent. (Schumm and Lichty 1965, pp. 118–119)

Table 8.2 The status of drainage basin variables during time spans of decreasing duration

Drainage basin variables	Status of variables during designated time spans		
	Cyclic	Graded	Steady
1. Time	Independent	Not relevant	Not relevant
2. Initial relief	Independent	*Not relevant*	*Not relevant*
3. Geology (lithology, structure)	Independent	Independent	Independent
4. Climate	*Independent*	Independent	Independent
5. Vegetation (type and density)	Dependent	Independent	Independent
6. Relief or volume of system above base level	Dependent	Independent	Independent
7. Hydrology (runoff and sediment yield per unit area within system)	Dependent	*Independent*	Independent
8. Drainage network morphology	Dependent	Dependent	Independent
9. Hillslope morphology	Dependent	Dependent	*Independent*
10. Hydrology (discharge of water and sediment from system)	Dependent	Dependent	Dependent

Source: after Schumm and Lichty 1965

Extract 2   Magnitude and frequency of forces
The relative importance in geomorphic processes of extreme or catastrophic events and more frequent events of smaller magnitude can be measured in terms of (1) the relative amounts of 'work' done on the landscape and (2) formation of specific features of the landscape.

For many processes, above the level of competence the rate of movement of material can be expressed as a power function of some stress, as, for example, shear stress. Because the frequency distribution of the magnitudes of natural events, such as floods, rainfalls and windspeeds, approximate log-normal distributions, the product of frequency and rate, a measure of the amount of work performed by events having different frequencies and magnitudes will attain a maximum. The frequency at which this maximum occurs provides a measure of the level at which the largest portion of the work is accomplished ... As the variability of the flow increases a larger percentage of the total load is carried by less frequent flows... The extreme [conditions] associated with infrequent events are compensated for by their rarity, and it is found that the greatest bulk of sediment is transported by more moderate events ... When stresses generated by infrequent events are incompetent to transport available materials, less frequent events of greater magnitude are obviously required. (Wolman and Miller 1960, p. 156)

Extract 3   Dynamic equilibrium in landscapes
Since the period 1890 to 1900 the theory of the geographic cycle of erosion has dominated the science of geomorphology and strongly influenced the theoretical skeleton of geology as a whole. Some of the principle assumptions in the theory are unrealistic. The concepts of the graded stream and of lateral planation, although based on reality, are misapplied in an evolutionary development, and it is unlikely that a landscape could evolve as indicated by the theory of the geographic cycle.

The concept of dynamic equilibrium provides a more reasonable basis for the interpretation of topographic forms in an erosionally graded landscape. According to this concept every slope and every channel in an erosional system is adjusted to every other. When the topography is in equilibrium and erosional energy remains the same all elements of the topography are downwasting at the same rate. Differences in relief and form may be explained in terms of spatial relations rather than in terms of evolutionary development through time. It is recognised however that erosional energy changes in space as well as time and that topographic forms evolve as energy changes. (Hack 1960, p. 80).

In Case Study 8.2 Schumm and Lichty's idea of causal relations being a function of the scale of time and space over which we consider landforms is simply a statement that geomorphologists have developed different theories to explain how landforms behave at different scales (Table 8.1). A variable such as vegetation is explained by a variable such as topography at one scale, while at another scale topography may be the explained variable rather than an explaining one, and so on. The danger in Schumm and Lichty's scheme is that it implies that it is in the nature of some variables to be dependent variables at one scale and independent variables at another. Whereas, whether they are explained at one scale or not depends on the existence of a theory which provides the explanation. Their scheme merely states the truism that theories explain variables and that only certain theories have been developed. Their ideas show how different theories are related to each other and do not in themselves constitute a scientific theory, since they exclude nothing and can never be falsified. Despite the mythical nature of their ideas their paper constituted something of a landmark in geomorphology. In terms of the history of the subject its significance lies in the fact that geomorphologists had failed to see that theories can be about different things and answer different questions and yet not be incompatible (Tinkler 1985). Acceptance of the historical approach is not incompatible with accepting the equilibrium approach since they deal with different problems at different scales. Schumm and Lichty resolved an issue which at the time was problematic but which in retrospect can be seen to be trivial.

Wolman and Miller's idea in the second extract of Case Study 8.2 can be expressed simply as the fact that the magnitude of the process events which control the form of particular landscape features is determined by the relative frequency of the magnitudes of events. Landforms, in other words, depend on particular combinations of the magnitudes and frequencies of events. Once again this is a truism and thus a myth. Whatever landscape feature is observed, it must have been formed by processes which had some magnitude and frequency. A similar point can be made about Hack's idea of dynamic equilibrium in Extract 3. His argument is that landscapes can always be considered as being in a state of dynamic equilibrium rather than in some stage of a process of change so that the form, physical properties and stress in landscapes could be considered to be mutually adjusted. The point of Hack's argument was that we do not need to erect theories of historical change in order to explain the present landscape. Present processes could account for what we see. This argument, however, does not exclude long-term studies of landscapes or show that there is no evidence for past events. It dismisses them on the grounds that they are not necessary for explanation of particular features of the landscape. This is another argument which cannot be falsified.

Accepting any of the ideas of Case Study 8.2 is not a scientific matter. We must not regard them either as theories or as providing an understanding of the world. Nevertheless such ideas are the basis for much of what goes on in science. We must avoid the mistake of dismissing them as unimportant since they have a role in the generation of theories. The ideas of Case Study 8.2 can best be regarded as concepts which promote questions. They cannot give rise to a problem through the failure of a prediction in the same way as the test of a theory. They can however lead to a problem in the way of saying: How would the matter look if we viewed it this way? The task of the scientist is then to translate a way of looking at the world into testable propositions.

Developing a theory

As we have seen, theories do not arise in isolation. The building of a theory is a response to a problem situation in which some prior theory has failed or expectation been unfulfilled. The scientist can react to such a situation and create a new theory in two ways. Either the old theory may be amended or a fundamentally new theory may be developed. In this section we will examine these two procedures in more detail.

Amending a theory

One way of amending an existing theory involves changing its premises. Vitousek and Reiners, for example (see Case Study 8.1), found that observations did not confirm their expectation about patterns of nutrient losses. Not all old forests lost substantially less nutrients in runoff than medium age stands. They therefore proposed that old forests are a mosaic of patches at different stages of succession, the younger of which are the source of higher nutrient losses as the forest as a whole matures. Thus they effectively introduced a new premise into their theory: that the theory only applies to mature ecosystems defined as uniform stands.

In adopting such an approach the logical process which Vitousek and Reiners go through is not so much one of theory building as one of theory protection by the amendment of premises. Despite the fact that the data of Fredrikson (1972) and much other work on forests contradicts their predictions, Vitousek and Reiners defended their theory against the accumulated evidence of mature systems being patchy, and releasing less nutrients than they gain, by considering that such patchy systems misrepresent truly mature systems. They assert that if a uniform mature system were observed it would show the nutrient losses they predict.

Vitousek and Reiners could however have reacted differently. They could, for example, have amended the structure of their theory more fundamentally. They could have said that since mature stands of vegetation are restricted spatially, there is no need to devise a theory in which the outputs per unit area are constant for a constant biomass, since for any confined area there may be inter-area transfers. A mature stand surrounded by immature stands may have extra inputs from herbivore droppings and windblow which allow it to maintain a greater output than could be expected from the inputs to surrounding mature forests. The testing of this theory would of course require different measurements from the testing of the previous theory of eventual net nutrient losses being zero, because its crucial point is about differences in inputs. Such a revision of the theory is achieved by considering input–output relations as part of a more general set of mass balance relations and recognizing that Odum's hypothetical relations are rather limited being only one of several possibilities. The problem of explaining the anomalous results then involves deducing what mechanisms could account for greater input to mature than to immature systems or what mechanisms of input might only apply to mature systems which could be reflected in increased output without altering the biomass.

An alternative tactic which Vitousek and Reiners might have used is speculating about mechanisms which could account for reduced losses from mature ecosystems without altering their biomass. Such amendments are achieved by the process of *reverse deduction*, that is, taking an effect and reasoning about the causal mechanisms which could have produced it. The knowledge which forms the basis of these speculations comes from accumulated experience of ecosystem studies. Considering any of the well-known mechanisms of nutrient cycling we can ask: Would this behave, under the conditions specified by the theory, in such a way as to account for the differences observed?

Vitousek and Reiners's study illustrates the process of putting forward a new idea whether it be an amendment to an existing one or something radically new. Being able to generate a new idea depends on familiarity with a repertoire of ideas, on experience with problem situations and on the ability to deduce the causes which could produce particular effects.

Creating new theories

From the analysis of theorizing so far it might be concluded that *new* theories can never be developed. However, there is the situation in which the hypothesis offered to explain a problem is radically different from the one which gave rise to it, and the term 'new' is used. In the attempt to develop new hypotheses or theories a factor which has been recognized as important (Gilbert 1896) is the range of *abstract ideas* which the scientist possesses and which he can call upon as analogies when faced with a problem of explanation (see 'Problem situations' earlier in this chapter). Such analogies need bear no relation to the original theory or problem situation. Feyerabend (1975) has argued that in terms of theory development 'anything goes' (see Chapter 6). In fact there is great value in presenting ideas which are:

> ... set forth simply as an outrage, to do violence to certain generally established views about the earth's behaviour that perhaps do not deserve to be regarded as established; and [are] set forth chiefly as a means of encouraging the contemplation of other possible behaviours ... deliberate enough to seek out just what conditions would make the outrage permissible and reasonable. (Davis 1926, pp. 467–468).

Although proposing such hypotheses is a serious business (Kennedy 1983), Feyerabend's advice, that new theories should be developed which disagree with accepted facts and theories, can be useful only in the sense that there must be *good reasons* for considering some accepted facts to be in error (see Chapter 6 and the discussion of Baker 1978). In other words it is unreasonable to deal with the problem of how to construct theories if any statement about any fact is to be taken seriously as a basis for theory development. This is not to say that a theory cannot or should not be invented to explain factual situations which are the opposite of what is accepted. Since we require all scientific theories to be testable, the invention even of absurd theories raises no problems. The difference between this and the method of 'anything goes' derives from the attitude of the proposer. There must be a bona fide attempt to advance knowledge. There is no merit in concluding that since all knowledge is uncertain there is no structure or method to knowing, and on this basis then lump the nonsensical, trivial and purely fanciful with the genuine ideas of scientists, however imperfect they may be.

How far can the scientist go in being bold or outrageous? It is possible to conduct a thought experiment involving confronting Aristotle with modern theories about the origin of springs and rivers (see Case Study 4.3). Because of the problems of incommensurability (see 'Incommensurability and relativism' – Chapter 4, and 'Rationalism vs relativism' – Chapter 7) we could easily imagine the shock of the conflict between our ideas and his. To the best of our knowledge our ideas are nearer to the truth than Aristotle's and yet how would he know? What future history of ideas lay before him he could not tell. On a similar basis we can argue: How can we judge outrageous hypotheses with which we are now faced? Could we not utter what, now to our ears, would seem absurd but would in time become accepted as a good theory? How are we to judge the boldness and originality of hypotheses?

The significance of this last question relates to the status of Feyerabend's proposal about scientific method. Beyond the bold and the outrageous he proposes that there should be an amorphous reservoir of ideas which are potentially usable as explanations, and any addition to this reservoir is within the scientist's domain. All the scientist has to do is choose from the repertoire. The response to this suggestion, and the answer to the question of boldness posed above, lies in our understanding of the development of science. Its development is through problem situations. The theory and the problem live together. New theories should explain more than those they are meant to replace but they still have them as parents. We cannot state a theory so bold or so outrageous that it lies totally beyond our understanding. And if we do we can certainly do nothing about testing it. Considering such a possibility is trivial.

The incommensurability of the two theories in the thought experiment does not imply that Aristotle's reaction to our theory as being absurd has the same status as us finding his theory absurd. It is because our theory, though it is far removed from his, nevertheless has grown out of it by the process of trial and error. The reason why he might find out theory absurd is because he is missing all the other stages of development which lie between us. The reason we would find his theory absurd is because we know the stages and can see its falsity content. It is this relationship between theories which contradicts Feyerabend's proposals.

The boldness of theory is more properly judged by its relation to the theory with which it competes and by its relation to the situation which led to the old theory's failure. A bold theory is one which corrects or contradicts the one it seeks to replace but which also accounts for the success of the old one. If the new theory passes the tests we put on it and the test where the old theory failed, then it is a better approximation to the truth, as in the case of the relation between the theories of continental drift and plate tectonics (see Case Study 4.1). Bolder theories have greater universality and greater precision. As universality increases so does the domain of possible falsification. Similarly, greater precision in prediction particularly through quantification increases the range of falsifying situations.

Any satisfactory new theory must always explain more than its predecessors and lead to new discoveries. It must not, in other words, be an ad hoc theory (see 'Falsification and ad hoc theories' – Chapter 3). Otherwise it would merely lead to explanations which are circular and untestable. In building new theories, therefore, the scientist should put forward bold conjectures which open up new domains of observations and which lead to new tests and new problems (Popper 1972b, p. 355).

Conclusion

Theories owe a great deal to other theories and to the myths that scientists hold. Amending old theories and developing new ones depend on these myths and other theories and may to some extent reflect the capacity to make analogies. Perhaps above all theorizing follows from persistent painstaking work and from the ability, produced by a critical attitude of mind, to recognize problem situations and to make bold, imaginative conjectures.

In considering the importance of boldness and imagination it would be a mistake to interpret these ideas to mean that any newcomer to a problem can, with a fresh view and an unconventional approach, make the brilliant leap to a new theory. No doubt such things occur but the image of sudden, inspired realization is somewhat over-romantic. Insights are not sudden isolated and unexplained mental leaps. They are the product of sustained interest and commitment to a problem, of repeated trial solutions and of constructive criticism.

It is an erroneous impression fostered by sensational popular biography, that scientific discovery is often made by inspiration . . . This is rarely the case. Even Archimedes' sudden inspiration in the bathtub; Descartes' geometrical discoveries in his bed; Darwin's flash of lucidity on reading a passage in Malthus; Kekule's vision of the closed carbon ring came to him on top of a London bus [sic]; and Einstein's brilliant solution of the Michelson puzzle in the patent office in Berne were not messages out of the blue. They were the final co-ordinations, by minds of genius of innumerable accumulated facts and impressions which lesser men could grasp only in their uncorrelated isolation, which – by them – were seen in entity and integrated into general principles. The scientist takes off from the manifold observations of predecessors, and shows his intelligence if any by his ability to discriminate between the important and the negligible, by selecting here and there the significant stepping stones that will lead across the difficulties to new understanding. The one who places the last stone and steps across to the *terra firma* of accomplished discovery gets all the credit. Only the initiated know and honor those whose patient integrity and devotion to exact observation have made the last step possible (Hans Zinsser, quoted in Harary 1971, p. 14).

Building theories about the world is not easy.

CHAPTER 9:
MODELLING

'. . . models are used to make predictions'

The role of models in science

A key element in the practice of science is the critical testing of theories. What is it that allows theories to be tested? It is that they can be used to make predictions against which observations can be measured. The success of the theory is then judged by how well its predictions compare with reality. However, judgement of theories requires knowledge not only of the theory and the observation but also of the means by which the prediction was obtained. Such means are also sources of uncertainty and need to be assessed in any test of a theory. This chapter is concerned with how models fit into the process of testing theories.

In this chapter we will refer to devices used to generate predictions as models. This definition is not one which is common in physical geography. Haggett and Chorley, for example, consider that a model is:

> . . . either a theory, a law, a hypothesis or a structured idea. It can be a role, a rela-
> tion or an equation. It can be a synthesis of data. Most important, from the
> geographical viewpoint, it can also include reasoning about the real world by
> means of translations in space (to give spatial models) or in time (to give historical
> models). (Haggett and Chorley 1967, pp. 21–22)

There is also a widely held idea that a model is an idealized or simplified representation of reality (Haggett and Chorley 1967; Holt-Jensen 1981; Thomas and
Huggett 1980).

We have chosen to give a different definition of a model because from the writings
of most physical geographers it is difficult to see what is not to be considered as a
model, or what it is specifically that models do. Theories and laws, hypotheses and so
on have identifiable and separate roles in science, and it adds nothing to our
understanding to call them models. The idea that models are simplifications of
reality is also misleading. A photograph may be a simplification of reality, but that
does not make it a model. All statements and representations are simplifications of
reality, and to describe some as 'models' conveys no information whatsoever. We
must ask, therefore: What are models and what distinctive role do they play in
science?

As we saw in Chapter 3, a theory can be tested by using deductive reasoning to
make a prediction. This reasoning was the *mechanism by which premises were used to
devise conclusions*. In contrast to the empty definitions of models described above, it is

this task to which the term modelling is more usefully applied. In general a model processes information to generate new information, that is:

> ... the model is any rule that generates outputs from inputs. (Overton 1977, p. 63)

In other words models are devices used to make predictions. Only when models are considered in this specific sense can their role in science be appreciated. In this chapter we will consider the different types of model, how models are constructed and how they are used in theory testing. First, however, it is necessary to examine the context in which models are used.

Case study 9.1: Models and science

Consider the work described below and in each case try to identify the models which they employ. How are the models used to provide a test of a theory? Are models used in the same way in both studies?

Model 1   Determination of the ground water component of peak discharge from the chemistry of total runoff

A major problem in hydrology is to determine which part of the storm runoff hydrograph peak is derived from groundwater and which part from direct runoff. Pinder and Jones (1969) considered this problem and suggested that if the chemistry of the two sources of runoff differs, then by using a simple equation to describe the mixing of the two sources it should be possible to estimate the contribution of each source to the total discharge of water. The model they devised was based on the idea that the total solute load of the waters leaving the catchment is the sum of the load dissolved in direct runoff plus the load in groundwater runoff. Solute load is the product of concentration and volume. Thus:

$$C_T Q_T = C_G Q_G + C_D Q_D$$

where C_T, C_G, and C_D are the concentration of solutes in the total, groundwater and direct runoff, respectively, and Q_T, Q_G, and Q_D, are the discharges of each of these three sources.

C_T and Q_T can be measured in a river and C_G and C_D can be measured by sampling direct runoff and groundwater flow. If both these components are known then only Q_G and Q_D need to be determined. Since total discharge is the sum of groundwater and direct runoff, Pinder and Jones argued that the groundwater discharge must be equal to:

$$Q_G = [(C_T - C_D)/(C_G - C_D)]/Q_T$$

Q_D is then found from $Q_D = Q_T - Q_G$

Pinder and Jones used these equations to estimate the groundwater component for the April Brook catchment in Nova Scotia. They used the chemistry of bicarbonates as a guide to the mixing of the two sources of groundwater. They found a peak discharge (Q_T) of 12 cfs, $C_T = 45.2$ ppm HCO_3^-, $C_G = 84$ ppm HCO_3, and $C_D = 18.8$ ppm HCO_3^-. Thus about 40% of peak discharge was provided by groundwater runoff.

Model 2   The influence of porefluid salinity on instability of sensitive marine clays

The susceptibility of post-glacial marine sediments to earthflows is well known. Rosenquist (1955) established from laboratory studies that marine clays laid down under saline conditions are more stable when their porefluid is saline than when the salts in the pores are leached out. Carson (1981) considered the implications which this result had for the stability of slopes in areas of marine clay. He argued as follows:

A logical geomorphological corollary [of Rosenquist's result] is therefore, that in an area of exposed marine clay sediment, large earthflows should not occur in those parts where the salinity remains high, but should occur, if other factors permit, where the sediment has been depleted of its porefluid ions. (p. 499).

Using a rapid method of estimating porefluid salinity, Carson showed that areas of large earthflows coincided with areas of low porefluid salinity. Earthflows were absent from intervening areas of higher salinity. The results also indicated that low porefluid salinities occurred only on areas of thin deposits resting on more permeable rock.

The desire to make predictions about the world is not only found amongst scientists. We make predictions all the time in our daily lives as we anticipate events and plan for the future. What distinguishes science, however, is the way in which predictions are used. The scientist employs a prediction as a means of critically testing a theory and as a guide to how he should modify his ideas about the world.

In order to bring out the special way in which scientists use predictions in their work, we will distinguish between the use of models in pure science and the way in which they are used in applied science or engineering. Like the engineer, the pure scientist can use a model to make a prediction. The aim of the scientist, however, is to compare this prediction with the real world as a means of evaluating the theory which gave rise to it. He can also use a model to test the internal consistency of a theory. In contrast, the engineer uses a model to make a prediction, but it is not used as a basis for a critical test of some theory, rather it is used as a basis for some action or a decision.

This distinction between the pure scientist and the engineer is of considerable significance in understanding the contribution which scientists make to knowledge. Reference will be made to it throughout this chapter although we are concerned principally with the scientific use of models and not with their use in applied science or engineering. In Case Study 9.1 the work of Pinder and Jones provides an example of the use of models in engineering. These workers were merely concerned with devising a way of estimating the groundwater and direct runoff components of discharge to aid in the prediction of future storm hydrographs. To do this they constructed a simple mathematical model. Their study included no independent test of their estimates of the various components from empirical data and thus no test of their method, irrespective of any test of their theory. They merely established that the method can be used.

In contrast to such use of models in engineering, the study of Carson (1981) in Case Study 9.1 illustrates the use of models in the more critical way typical of the pure sciences. Even though Carson was concerned with an engineering problem in the stability of slopes, his study used a fundamentally different approach. He *argues* why he should find a particular pattern of instability and establishes by means of observation that his ideas are upheld. To do this Carson utilized a simple logical model to make a prediction.

In making a distinction between the use of models in a pure and an applied sense there is an implied criticism of the so-called scientific use of models for prediction which is not directed at any critical evaluation of ideas. There is no implied criticism of the use of models in applied work. The different attitudes of the scientist and engineer are conditioned by their respective aims and are valid in context. The engineer may not bother to test the theory on which the prediction is based, because it has been used successfully in the past. However, it is clear that even though the engineer may not want to test a theory, critical testing is required at some stage if any rational basis for action is to be found. This rational basis is provided in tests of theories by pure science and it is with the use of models in this context that we will be concerned here.

Models in testing

For some theories models can be developed which will test the theory without any reference to observations of the world. An example of this is provided by May's analysis of model ecosystems (May 1975, 1981). May was interested in testing theories about the relations between ecosystem diversity and stability. It had been suggested (for a review see Goodman 1975) that more diverse ecosystems are more stable because there are more interconnections between elements of the ecosystems, so that the system as a whole is able to withstand disturbance better than a simpler ecosystem with less connectivity. The problem in testing the so-called diversity–stability hypothesis was that there was no unambiguous way of measuring complexity or stability. Therefore, in order to test the hypothesis, May built an abstract, computer model of an ecosystem. This system could then be used to investigate the way in which complexity affected its stability. By simulating the behaviour of ecosystems which differed in their complexity, he was able to show that complexity and stability were not related. He speculated that in the case of real ecosystems, in which stability and complexity were associated, their behaviour must be due to evolutionary development rather than to any feature of their structure.

May's study represented a major advance in ecology because it refuted what many considered to be a key ecological principle. In the present context it illustrates the value of what might be considered as 'thought experiments' in the testing of theories. Simply by building a model based on how he thought ecosystems were structured, and by analysing the logical consequence of these assumptions through the model, May tested a theory without any reference to the real world. He thereby showed that the diversity–stability hypothesis was inconsistent with certain other accepted ideas in ecology, and this allowed him to reject it.

A difference between a prediction and the criterion against which it is set can arise for a number of reasons. The first, as we have seen, is that the theory is wrong. This can only be shown unambiguously when the theory alone is the criterion against which its model output is judged. There are other criteria, but before they are described the structure of the testing process must be clarified in order to comprehend the roles and relations of its various elements. The basic structure is shown by May's analysis. He tested the consequences of a theory, translated through a model, against the consequences of the theory, translated through deduction. In other words the test situation comes down to the logical problem of consistency. As we shall see, this same structure can be considered to apply to *all* test situations although there are many sources of uncertainty apart from theories.

Consider the situation in which a theory is tested by comparing the predictions from a model with observations of the world. What are compared are the consequences of a theory translated through a model and the consequences translated through observations. This is so because the observations are taken in relation to the theory, that is, they are taken to test it. They follow logically from the theory in a different manner but no less than the model (for a fuller examination of this relation see Chapters 10 and 11). The problem in testing then becomes: To which element of the testing process should any discrepancy be ascribed, the theory, the observations or the model?

The observations can lead to a discrepancy in two ways: either because the bridge principles of the theory (see Chapter 1) are in error or because the taking of the observations was in error. These problems of error in measurement and experimentation are dealt with in Chapters 11 and 12.

The model may lead to a discrepancy for one of three reasons: first, because it contains a random variable; second, because the model is not fully specified by the theory, and requires calibration using empirically derived information, which may be subject to error; third, because the components of the model may be indeterminate in their behaviour. These characteristics provide the basis for a classification of models as they are used in science, that is, in the testing process. The classification is based on sources of error. There are three criteria:

1. whether the model is deterministic or stochastic

2. whether the model is fully or partially specified
3. whether the model is a hardware or a software model

A fuller classification of models is presented in Figure 9.1. We will examine the three criteria in turn.

1. Whether the model is deterministic or stochastic

For some models a single input produces a single output. Such models are called deterministic models, and are illustrated by the model of Pinder and Jones, in Case Study 9.1. In contrast there are models which have two or more possible outputs for a given input. Such models are used when part of a theory about the world may be that its behaviour cannot be described simply by the mechanisms contained within the theory. The existence is recognized of other unknown effects which are considered to produce random fluctuations in an observed variable. Models are built to incorporate a random element or a random part of their structure to simulate this effect. Such models are known as stochastic models. With models of this type there may be a discrepancy between a model prediction and an observation which may be due to the random element. A simple kind of model which includes a random element is a Markov Chain in which the state of a variable depends on a previous value and a random factor. Such models have been used successfully to simulate drainage networks (Shreve 1975). In general, however, these models require the calibration of the random element, as in the modelling of the succession of geological strata in bore hole records (Harbaugh and Bonham-Carter 1970), or of forest succession (Horn 1981). Such calibration involves another source of uncertainty and therefore another criterion in the classification of models.

2. Whether the model is fully or partially specified

A simple example of a fully specified model is one of Newton's laws of motion:

$$s = ut + \tfrac{1}{2}ft^2$$

where s is the distance travelled by an object after time t having had a starting velocity u and been subject to a uniform acceleration of f. Given values of these variables, v, f and t, we can predict s.

An example of a model which is not fully specified by a theory is the Rational Formula (Dunne and Leopold 1979):

$$Q = C.I.A.$$

which predicts the peak discharge Q from small areas where I is rainfall intensity and A the area. C is a constant determined empirically from previously measured values of Q, I and A. Q cannot be predicted directly from a theory since a prediction involves

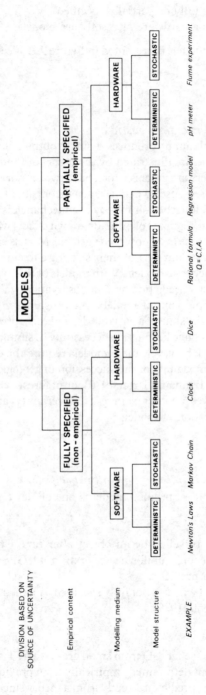

Figure 9.1 Classification of models as they are used in science

using an empirical value C. The significance of this is that, in testing a predicted value of Q against an actual value, a discrepancy may arise from an inadequate estimate of the value of C, rather than as a result of the model or theory.

3. Whether the model is a hardware or a software model

The third source of uncertainty, used as a criterion for classifying models, is whether the model is a hardware or a software model. The term software is used to describe those models which exist as abstractions and which can be represented symbolically. Often the word software is used to describe computer programmes which may or may not be models. Nevertheless, the term is employed here partly to contrast with the term hardware, as in computing, and partly to show that these models can be represented abstractly. The term 'conceptual' is often used to describe such models, but it is not sufficiently precise. Although the term mathematical is technically more correct, too often it is misinterpreted or associated with particular symbolic procedures.

The simplest type of software model is represented by the covering law model. It is one of the logically valid forms of reasoning. An example of a theory, or premise, used in the covering law model is: When it rains the river will rise. The covering law model then allows us to predict the outcome from any observation:

If it has rained the model predicts the river will rise.
If it has not rained the model predicts nothing.
If the river has risen the model predicts nothing.
If the river has not risen the model predicts that it has not rained.

Knowing this set of situations then allows us to test the theory. We can only say anything about the theory with the first and last situations, since only in these does the model predict. If we observe that it has rained and the river has risen, then we can say that observations have not refuted our theory. It may be true or it may not since the river could have risen for other reasons. If, on the other hand, it has rained and the river is not observed to rise, then this refutes our theory. This is the *critical test situation*. In both the other situations, when it has not rained, we can come to no firm conclusions. Thus the covering law model provides predictions, or outputs from our inputs, and specifies the way in which a theory is to be tested.

So far it has been assumed that the models that are used to provide predictions are abstract. However, models can be used which are composed of real objects, which are meant to stand in for the objects contained in the theory in the same way as abstractions do in software models. Such models are called hardware models. Like software models they can be used to generate output from input.

Consider the problem of testing a theory of sediment transport rates in rivers. A hardware model of a river channel consisting of a sediment-filled flume and water can be built, and this can be used to predict sediment transport or flow under different conditions (Grass 1970, 1971). These outputs can then be compared with observations in real channels. A corresponding software model would be a set of equations describing sediment transport (Bagnold 1966).

When comparing predictions of the model channel with observations of real channels, we are considering the effects not only of the theory, which includes the effects of scale, and of the observations but also of all the unspecified aspects of behaviour of the model. In abstract models these can be represented by specified random functions but in the hardware models they are not. Indeed, one of the reasons for using hardware models is the hope that they may reproduce the complex effects of a real system which cannot be specified in an abstract model.

Another type of model which can be considered along with hardware models is what Chorley (1967) describes as natural analogue systems. These are real objects or events which are believed to behave in the same way as the object or event of interest. Thus in forecasting next week's weather we can use past meteorological situations similar to today's. Next week's weather will then be taken to be like that which followed the previous situation. However, the crude use of such analogues amounts to no more than an attempt at inductive reasoning. That is, saying that on the basis of repeated observations the future will be like the past. Nevertheless, the use of analogues is a perfectly acceptable procedure for forecasting or prediction when the objective is to decide whether it will rain or not next week. But these things are not part of science.

The term analogue models is often used to describe such things as electrical circuits which are used to predict the behaviour of, for example, hydrological systems. Their workings or mechanisms are specified by the circuits used in their construction and they cannot be regarded as being the same as hardware models, as the term is employed here. Rather, they are the same as software models in that their functions can be specified by theory or determined empirically. They have some advantages over abstract mathematical models in applications to particular problems to which digital computers only provide approximate solutions. These usually involve complicated differential equations. However, their use in physical geography is minimal.

Empirical development of models

One of the classes of models described above is of those which are partially specified. Their use in science depends upon calibration. That is, the fitting of empirically determined constants or coefficients in order that they may be used to predict. The use of such models is more common in engineering, but it is also a valid part of the scientific development of models since not all theories are sufficiently well formulated to specify the full structure which a model should take. An example of the empirical development of a model which falls across the boundary between the two approaches is given in Case Study 9.2.

Case Study 9.2: Catchment modelling
Predicting flows from ungauged catchments is a central problem in hydrology. It can be approached using simple empirical relations which make little use of established theory about runoff generation. In contrast, Nash and Sutcliffe (1970) suggested the use of runoff models based on physical laws which incorporate the characteristics of a basin in the structure of the model. They considered that if a model is to help in the understanding of the process of converting of rainfall into runoff, it is essential to obtain some guide to the relative significance of the parts of the model and the accuracy of empirically derived parameter values. The problem is that of the conflict between using sufficient parameters to describe a basin accurately and the need to determine their values, which can only be done satisfactorily when they are small in number. The problem cannot be resolved completely. Nash and Sutcliffe recommend the isolation of operations within the model and then separate fitting of parameter values by an automatic process of optimization, as follows:

> ... automatic optimisation [uses] an index of disagreement between the observed and computed discharges ...
>
> $$F^2 = (q'-q)^2$$
>
> where F^2 is the index of disagreement and q' and q are the observed and computed discharges at corresponding times.
>
> The quantity F^2 is a function of the parameter space and ... of the input and output. Optimisation involves finding the values of the parameters which minimise F^2. This can be done by a 'steepest descent' method, or a search can be conducted ... by moving parallel to the parameter axes ... [as shown in Figure 9.2 which is of] a cross-section through an optimum parallel to one (X) axis ... and of the same cross-section through dF^2/dX. The intersection with the X-axis is at Xopt. The higher the angle of interception, the better is the definition of Xopt.
>
> If substantial dependence exists between two or more parameters this [method]

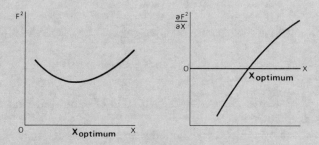

Figure 9.2 Optimization of F^2: single parameter case

is not sufficient. In [Figure 9.3] the dependence of F^2 on X and Y is indicated by a set of contours. Dependence between X and Y is indicated by the valley in the surface roughly along X+Y = constant.

The optimum values Xopt and Yopt may indeed be found . . . [but] . . . while a function of Xopt and Yopt is well defined, the separate values are not. The occurrence of such a relationship could be discovered by taking the second derivative in all directions from the minimum point of F^2.(pp. 288–289)

The development of catchment models of runoff in the manner described by Nash and Sutcliffe is presented in the companion papers by O'Connell et al. (1970) and Mandeville et al. (1970). They show that optimum values of model parameters can be

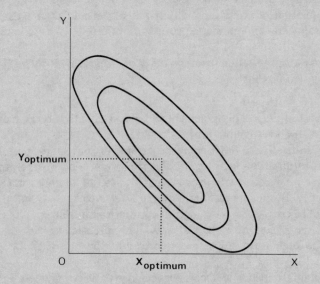

Figure 9.3 Optimization of F^2: double parameter case

found which allow a choice between competing models and successful predictions to be made.

Consider the material of the Case Study and if possible of the papers on which it is based.

How would the scientist respond to the situation of optimized models not producing good predictions?
If the scientist achieved a good prediction with such a model, how would he respond to this situation?
Can scientists build simple models of complex systems?

It is with models such as the one described in Case Study 9.2 that the greatest difficulties are presented in testing theories. The model is of different parts of the runoff process and contains several components each with empirically derived parameters. As Nash and Sutcliffe point out, the problems of deciding where the sources of error are cannot be settled completely. But, by a gradual process of optimization and model restructuring, it may be possible to achieve predictions which are sufficiently close to reality to provide the basis for action. As we have said, however, this is not a problem of science, and it would seem that such development of models is at present outside of science, since it hardly provides a method whereby theories are seriously tested. The main thrust of the work of Nash and Sutcliffe and their co-workers is towards testing the 'known physical laws'. You might like to try Exercise 9.1.

Judging predictions

Having considered the various types of model in terms of their sources of uncertainty, we must now examine in more detail how they can be used to provide a test of some theory. Conceptually, the process of testing is very simple. The model is used to make some prediction and the scientist must then make a judgement about whether the prediction is sufficiently close to what is expected in order to retain the theory, or sufficiently different in order to reject it. As we have seen, however, the procedure may be difficult to follow since an inconsistency between the prediction from a model and an expectation may arise in several ways apart from the inadequacy of some theory.

Testing involves making a *judgement* about the predictions from a model. The predictions can be either qualitative or quantitative irrespective of whether the model is qualitative or quantitative. Examples of the different combinations of models and predictions are presented in Table 9.1. In making a judgement, however, the important distinction is in the nature of the prediction.

Judging a qualitative prediction, for example that volcanos are present or that a river has a particular channel pattern (Table 9.1), involves specifying in advance that some object or event belongs to a particular class. Knowing that an object is a volcano, that is, it belongs to that class of objects which have the characteristics which we take to define volcano, is not usually a problem. However, it is not nearly so straightforward a task to decide whether a river channel is braided or not. It is up to the scientist to judge if the observation belongs to the class of interest. He can make independent observations of the object to see if they are consistent with other attributes which he might expect from the theory. Beyond that, the acceptance of the prediction as correct remains a matter of judgement.

With quantitative predictions the problem of judgement depends on whether the results can be fitted into a statistical model. Chapter 12 on experimentation describes some of the problems of interpreting experiments. In particular it describes the structure of statistical arguments and the problems of using levels of significance. It is sufficient to note here that, where there is a quantitative prediction from a statistical

Table 9.1 The nature of predictions from models

Model	Prediction	Example
Quantitative	Quantitative	$s = ut + \tfrac{1}{2}ft^2$ where s = distance travelled after time t u = initial velocity f = acceleration t = time
Quantitative	Qualitative	For discharge Q, if channel slope S is greater than $0.6Q-0.44$ then the channel will be braided rather than meandering (after Leopold et al. 1964, p. 293).
Qualitative	Quantitative	Presence of green crystals in a breathalyser test indicates blood alcohol levels above 80 mg/l.
Qualitative	Qualitative	Volcanos occur along plate margins.

model, the probability of achieving that outcome can be stated under the assumption that the theory is false. If the probability is less than a chosen value, the level of significance, then the assumption of the theory being false is rejected. Such interpretation still leaves the problem of judging the level of significance. Nevertheless, the existence of a concrete value of the probability of a result provides a yardstick for judgement.

With quantitative predictions which do not involve a statistical model, there are no probability statements on which to base a judgement. The problem of judging the success of such a prediction can be seen by considering as an example Sheilds model for particle entrainment, which relates clast movement to dimensionless shear stress and channel bed roughness (see Figure 9.4 after Miller et al. 1977, quoted in Leeder

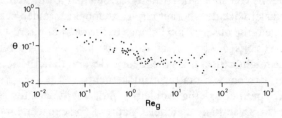

Figure 9.4 Dimensionless applied shear stress (θ) against Reynolds Number (Re_g) for initiation of grain movement
Source: after Miller et al. 1977

1982). The judgement of predictive success is based on a visual interpretation of the pattern of results. There is some notion that the pattern, predicted by the model for the threshold dimensionless shear stress at which particles move for any given channel roughness, is sufficient to compare visually with an empirically derived pattern, since the chance of achieving the pattern must be very remote. The problem is akin to judging qualitative models and significance levels.

Conclusion

Physical geographers have used the term model to describe so many different things that the word has come to have little meaning. The object here is to encourage the use of a specific meaning which has some relevance to how models are used in science. Models are only usefully considered to be the mechanisms by which predictions are made.

The significance of models derives from their role in the testing of theories. Testing is concerned principally with identifying the failure of theories to predict what is observed. A failure to match theory to observation may be due to factors other than the inadequacy of a theory. In particular, modelling may produce errors.

Thus the role and status of models in science can be considered in terms of the errors they are likely to produce since dealing with models in the test situation involves identifying sources of error. This reasoning provides the basis for a three-way classification of models according to their sources of error, which are: their empirical content, the presence of random elements, and the nature of the modelling mechanism.

CHAPTER 10:
NAMING AND CLASSIFYING

'. . . classification is fossilisation!'

Introduction

It may appear unusual that the consideration of classification follows that of theorizing and modelling. Naming and classifying are often represented as a preliminary to any scientific work. Abler, Adams and Gould, for example, write:

> The purpose of classification is to give order to the things we see so that we may learn more about them. If we did not classify, for example, we could not give names to constructs like 'cabbage', 'Saturday' or 'rabbit'. If we did not classify objects and experiences in commonly accepted terms we could not transmit information because no-one would understand what we were talking about. Nor could we make inductive generalisations because each event could be understood only as a unique experience. Thus language itself is an important classificatory device invented by the human mind.
>
> Because it helps us form hypotheses and guides further investigations, classification is the first big step taken by most sciences. Some scientists go so far as to argue that the state of classification in a science is a measure of its level of development (1971, p. 149)

Thus in postponing a consideration of naming and classifying until topics such as theorizing and modelling have been discussed, it may seem that an essential step in the scientific process has been missed. Such an ordering has to be deliberate in order to emphasize that the scientific process does not proceed according to some linear sequence of steps, and that emphasis should be placed on theory and testing not on classifying, which, as we shall see, is of limited significance in science. Knowledge does not grow by accumulating observations but by critical testing of ideas. Observations are guided by theories about the world and are made with particular objectives. They are not merely the passive responses of our senses. Recognizing the similarities between objects, grouping them in classes and giving names to them are activities undertaken in an altogether different way from that envisaged in the quotation above and in the classical view. Naming and classifying are only undertaken in the context of some theory, and the status of names and classes can only be judged in relation to that theory. The relationships between names and theories will be examined by considering the process of classification, the problem of certainty in naming, and the properties of classes in science.

Classification

Classification is the process of assigning the properties of a class to an object or situation, where a class is a set of things which has been defined by the possession of particular characteristics. It is one form of the general process of measurement (see Chapter 11). Thus we may have the class of all objects which are mudslides, or of all drainage patterns, or of all trees, and so on. Each is defined by its own class characteristics. Assigning a thing to the class 'mudslide' means that we recognize it as having particular properties. One implication of this view of naming and classifying concerns the structure and role of definitions in science.

Consider the definition:

> Mudslides are relatively slow-moving lobate or elongate masses of softened argillaceous debris which advance chiefly by sliding on discrete boundary shear surfaces.

Such a definition should be read from *back* to *front*, or from right to left in English. Though the definition apparently starts with a defining formula ('mudslides are . . .') and ends with a label for it ('. . . relatively slow-moving. . .'), in fact the scientific view would be that the definition is constructed in response to the question, 'What shall we call a slow moving or elongate. . .?' rather than to the question, 'What is a mudslide?'. 'What is?' questions are aimed at establishing the *essences* of things and as such have no place in science, for there are no essential landforms, or other things, waiting to be discovered. Mudslides, like all other objects, are recognized by the characteristics we have chosen. We can see that names do not play a very important role in science, for they can always be replaced by longer expressions (Popper 1962, II, p. 14). What are important are these long expressions. It is these which represent theories and are the bases on which objects are recognized. This process can be extended to the process of multivariate classification. The idea that multivariate techniques can be used to recognize the natural or the fundamental groupings of things is misguided unless the attributes included in the analysis are specified by some prior theory. The grouping of things in the absence of a theory can only be arbitrary. Multivariate classification is not some filter through which we can pass the world. To use it in such a way is no more than an attempt at inductive reasoning. Its use is valid only when it is a device by which we apply and test theories.

Properties of classes

A class is a group of things or situations which share one or more characteristics prescribed to determine membership of the class. They are often labelled by a name or a phrase. There are many types of classes, but in science we are generally interested in universal classes. They are universal in the sense that there are no limits to the membership of a class though in reality it contains a finite number. Our scientific propositions are not about some members of a class, some mudslides or some trees, but about all mudslides or all trees. We can of course make scientific statements about some trees, say oak trees, but then we have defined a new class with its own particular characteristics. However, we would not make a statement about oak trees which applied only to some oak trees.

Even though we deal with universals, classes can be empty or single valued. A class such as '800 kph surface winds' may never have been recorded or even existed. However we can still make statements about such empty classes and generate valid, testable deductions. A class such as 'the highest mountain in the world' which has only a single member is similarly one about which we can make testable statements. In contrast, there are some classes about which we cannot make testable statements, an example of which is the class 'fairies'. If we are interested in saying something scientific about a member of a class we can define and measure its attributes. These measurements and statements about them are intersubjectively testable. Consider the case study below and try to identify classes which have and classes which do not have this property.

Case Study 10.1: Community structure, population control and competition

Hairston et al. (1960) attempted to make generalizations about broad groups of organisms and were particularly concerned with trophic levels. Their conclusions can be summarized as follows. (1) Depletion of green plants by herbivores is rare. (2) Rare instances of depletion do occur where herbivores are . . . protected from predators. (Thus) 'the usual condition is for populations of herbivores not to be limited by their food supply.' (3) Herbivores are not limited by the weather. (4) It follows that herbivore control is (by) predation. (5) 'Predators . . . must limit their own resources and as a group they must be food limited.' (Murdoch 1966, p. 219)

Murdoch analysed the paper by Hairston et al. and questioned whether we can subject the predictions made by it to test by observation or experiment.

Three kinds of study have been considered as possible tests: studies on (1) single populations, (2) groups of populations, or (3) trophic levels. It has been pointed out by Hairston (1960) that the idea cannot be tested with reference to single instances of the idea. Thus it is agreed that the first type of test is unacceptable. Secondly, there is the suggestion that the conclusions are true of herbivores, carnivores, etc. in general. Murdoch writes:

> How do we establish such a general pattern or trend? Does this require that say, 75% or 80% of all herbivores, or of all herbivore species, are predator limited; or conversely, how many contrary instances are necessary to refute this hypothesis? Clearly the idea as stated in the paper provides no criteria for judging it in this way, and the second type of test is also not appropriate . . . Finally, we are left with

the possibility of testing if the conclusions hold for trophic levels as a whole and this requires an examination of the nature of trophic levels. Unlike populations, trophic levels are ill defined and have no distinct lateral limits; in addition tens of thousands of insect species, for example, live in more than one trophic level either simultaneously or at different stages in their life histories. Thus trophic levels exist only as abstractions, and unlike populations they have no empirically measurable properties or parameters. (1966, p. 223)

Consider Murdoch's analysis: identify the classes of things dealt with above.

Which are universal and which are not?

On what basis could Hairston et al. (1960) argue that single instances do not refute their theory?

On what basis did Hairston et al. (1960) classify organisms? Do you think their classification is valid?

Although the type of class represented by things such as ecosystems and trophic levels are dealt with in science, they raise particular problems. We can make statements about them but they are unlike classes such as plants in one important respect. As Murdoch states, ecosystems have no properties that can be established with any degree of certainty. We could delineate a patch of ground but not an ecosystem. How do we know where ecosystems begin and end? To answer this we must ask ourselves, 'What is not an ecosystem?', and the answer is, 'Everything can be an ecosystem'. We can easily say what is not a river when dealing with rivers, but it is not possible to identify what is not an ecosystem or a landscape of equilibrium or a non-graded river. Such things are concepts, they are non-testable. To the extent that science deals with testable statements, it deals only with those classes of objects which can be observed and about which such statements can be made. The idea of ecosystems and trophic levels is a myth which forms part of the background to theories and hypotheses. A classification based on a myth, cannot produce classes which are identifiable, real and observable. Try Exercise 10.1.

The certainty of naming
Even though science deals with real, measurable objects there is nothing certain in our knowledge about them. Such uncertainty arises from the conjectural nature of all knowledge. Nevertheless, we act in the belief that theories and observations are sufficiently certain for us to use them, but there is always the problem of certainty, that is, of whether these beliefs are well founded.

In making statements about the things we observe there are two problems relating to the issue of certainty. To illustrate the first, imagine that a scientist had a theory about coastal mudslides, that their density along any particular stretch of coast depended on the susceptibility to expansion of the clay minerals of coastal cliffs and the height of the groundwater table. In relation to the testing of this theory he might state that there are, say, 20 mudslides per kilometre along the Lias Clay coast of Dorset, England, as compared to 5 per kilometre along the boulder clay cliffs of Yorkshire, England. Someone might ask, 'Are you sure?' The scientist might then double-check his figures and confirm the original statement. However, if it was important, it might be insisted that he make even more certain what the figure was by going back into the field. And if his reputation depended on it, he might in fact remeasure and excavate so as to make sure that the features that he was dealing with were mudslides and did number 20 or 5 per kilometre. What this illustrates is that there is no absolute certainty about factual statements. It also illustrates that science does not demand absolute certainty.

If someone was prepared to accept the figure of 20, at first, without any criticism, then it was presumably because, in testing the theory, small errors in the value of the number are unimportant and so it is not necessary to have the exact figure. There was enough certainty in the first value which was reported in good faith. Whereas if, say, the value of 20 appeared anomalous, then it would have been regarded as more important and might have needed to be checked. The degree of certainty achieved or accepted depends on the problem in hand. 'We act upon beliefs ... everything depends on the importance attached to the truth or falsity of the belief' (Popper 1972b, p. 78).

There is a second aspect to the problem of certainty which relates to naming and classification. The scientist might be asked, for example, 'How do you know that the objects you were measuring were in fact mudslides?' In response he might say that he used the definition of Hutchinson and Bhandari (1971). However, this might not be an adequate definition for the purposes of testing his theory. There might be several types of 'slow-moving lobate or elongate masses of argillaceous debris which advance by sliding along discrete shear planes' which are affected by different mechanisms. Such mechanisms could be controlled by vegetation cover or clay mineralogy, which cause mudslides to behave in different ways. If the theory about mudslides was about their behaviour in relation to a factor not in their definition or in the way they were measured, and the scientist observes mudslides irrespective of these controls, then there would be a problem of explanation which would have arisen simply as a consequence of definition. This situation leads to the problem of equifinality (Chorley 1962). This is the idea that the same landform, thing or situation, can be produced by more than one means; in other words that two different causes can produce the same effect. Such ideas have been widely employed in physical geography to explain the apparent similarity of things, such as tors, which are produced in different ways (Gerrard 1978, 1984). The problem arises because they have been defined and measured using a limited set of criteria, say their morphology, and yet theories about them relate to other attributes. It is in these other attributes, for example those related to the processes of formation by tropical weathering (Linton 1955) or periglacial erosion (Palmer and Radley 1961), that the differences between the causal mechanisms may be reflected. Thus in testing a theory about objects or events, we should classify or measure them on the basis of attributes which are contained within the theory. We cannot develop theories about objects which are defined and measured in ways irrelevant to the theory.

A corollary of this argument about naming is that, as knowledge grows, and theories are replaced or improved, so definitions, names and classifications change. A name can come to mean more than it did in the past as our knowledge of things grows. Classes are subdivided or joined as we come to learn that what we regarded as the same are in fact different, or vice versa. Naming and classification are continuing processes of science. To approach them as if they are the first steps in establishing the truth is counterproductive. As Sir Mortimer Wheeler put it:

> ... classification is fossilisation! It is the petrification of all organic thinking and that timeless curiosity from which all discoveries spring. (Wheeler 1967, p. iii)

Summary

Science deals with universal classes, which are defined and named on the basis of theories. The attempt to classify or name in advance of theory is meaningless. Naming is the simplest form of measurement and to the extent that it is based on theory it contains a level of uncertainty. The scientist can, however, name or classify with whatever degree of certainty desired; it all depends on the problem in hand. If definitions or names are unsatisfactory then the scientist can revise them in relation to the problem which renders them inadequate. A particular problem arises when definitions and classifications do not relate to the theory being dealt with. Then there is a danger that scientific argument will degenerate into argument about the meaning of words. Naming and classification must be seen as a rather unimportant part of science as reflected in the fact that they are continually subject to change as theories develop.

CHAPTER 11:
MEASUREMENT

'There are many kinds of errors'

Introduction
Many regard measurement as the hallmark of science. They are wrong. Measurement is merely a by-product of the need to test and refine theories. This is not to say that measurement is unimportant. On the contrary, an understanding of the process of measurement has implications for the conclusions which the scientist can draw from observations of the world because the information which observations convey depends on the properties of numbers, on the constraints of statistical theory and on the errors measurements may contain. In this chapter we will consider the place of measurement in science.

Measurement
Measurement is often taken to be the process of representing some property of an object quantitatively. However, assigning an object to a class, although it is non-quantitative, is nevertheless a form of measurement. In this simplest measurement, that is, classification or naming, we might, for example, assign the properties of the class 'river' to an object and call it a river. The properties of the class could be taken for the purpose of discussion to include:

– open channels containing flowing water
– the channel has been formed by the water flowing in it
– the water has unidirectional flow.

The object we now call river is understood to have these properties. A class and its properties are abstract things. The significance of this is that as a purely abstract process we can deduce the consequences of a thing called a river having these properties without reference to the actual physical object itself. The properties of the class are part of world 3 which stand in for the real object. Of particular significance are the properties of classes which are represented by numbers. Consider that we have the class of objects called 'river' and the class of objects of width 10m. If we have an object which looks like (fits the definition of) a river and has width 10m it then belongs to both classes and is called a 10m wide river. This is the type of measurement which is called *quantification*. This is the process of assigning numbers to attributes of objects or systems according to certain rules. This is done in the belief that these numbers

164

represent the objects and how they behave in the same way as naming an object represents it. The significance of quantification lies in the fact that having assigned to the object the properties of a numerical class we have given the object the properties of a particular number. If we manipulate the number, for example by adding it to another and dividing by two to find the mean, we also believe that this mean represents some aspect of the properties of sets of objects. Thus in assigning numbers to objects we have assumed that algebraic or geometric manipulation represent relations between objects, properties of groups or structures of systems. The significance of quantification is that it allows the powerful tools of mathematics to be used. It allows more specific assertions to be made, bolder theories to be developed and more critical tests to be constructed. Not all forms of measurement, however, convey the same amount of information. There are various properties which measurements possess which condition how they can be used. In order to understand the role and value of measurement in science these various measurement properties must be examined in detail.

Measurement properties

There are four different procedures for measurement which provide four *number scales*: nominal, ordinal, interval and ratio (see Table 11.1). The properties of each increase in that order with each successive scale having the properties of the one before and one additional. These properties are: identity, order, additivity of differences and additivity of numbers.

Measurements at the nominal scale have only the property of identity. That is, the objects measured have been assigned to classes with no numerical properties other than the frequency with which they occur. Examples of such measures are names of objects or attributes, e.g. colour, landform, plant species morphogenetic region, whether an object is or is not a river.

Table 11.1 Properties of numbers

Scale	Property	Examples
Nominal	$\square = \square$	Colour, Soil type
Ordinal	$a > b, b > c$ $a > c$	Moh's Scale of Hardness Lichen Index of Air Pollution
Interval	$20 - 10 = 70 - 60 = 10$	Temperature, Date of Calendar
Ratio	$\dfrac{20}{10} = \dfrac{50}{25} = 2$	Length, Velocity, Weight

Ordinal scale measures have the additional numerical property of being in a class which has a relative size to other classes. That is, we can order the objects into classes whose members differ in magnitude. The intervals between classes are not generally known. Examples of such measures are indices of one attribute based on another measure, e.g. Moh's scale of hardness, pictorially based indices of roundness of sand grains or lichen indices of air pollution. Thus in the SO_2 pollution scale described by Hawksworth and Rose (1976), which is based on the occurrence of lichen communities at a given pollution load, we can say that an area with pollution index 2 is more polluted than an area with index 1 and less polluted than one with index 3, but we cannot say that 2 is twice the pollution of 1 or .667 of the pollution of 3. Nor can we say that if we add the pollution of the three areas together it will be the same as pollution level 6.

In addition to identity and order, interval scale measures also have the property of additivity of differences. This arises as a consequence of the fact that at this scale the *intervals* between numerical units are equal, and so the numbers can be added and subtracted. The only common measures which are in interval scale are temperature (°F, °C) and the date of the calendar. We can say, for example, that $20°C - 10°C = 60°C - 50°C = 10°C$. The differences between units are equal and this equation applies whatever the temperature scale used. What we cannot say, however, is that 20°C is twice 10°C which is readily seen if we convert to °F. Only ratio scale numbers have such a property.

Of the four scales of measurement, the ratio scale contains most information about an object. This arises as a consequence of the fact that the numbers themselves can be added. Thus the numbers can be divided and multiplied. A corollary of this is that the scales have an absolute zero. Examples of such measures are length, weight, age and area. We can say that a river of length 20km is twice as long as one of length 10km and the same is true even if we convert it to different units. We can also deal with ratios and make statements such as: the ratio between a river length of 20km and one of 10km is the same as that between river lengths of 100km and 50km, or $20/10 = 100/50 = 2$. Ratio scale measures contain the largest amount of information about the objects they represent. In order to get familiar with the different number scales you might try Exercise 11.1.

Measurement and theory

It has been argued in Chapters 1 and 2 that measurement and theory cannot be considered in isolation. There are two aspects to the relationship between them. The first is what lies behind both the reasons for the taking of measurement and their interpretation. We can describe this aspect of the relation as being concerned with *scientific* theory. In the second aspect are the theories which describe the relationships between measurements. These are the theories of *statistics*. The relation of measurements to each type of theory will be discussed separately.

1. Measurement and scientific theory

Measurements are taken for particular reasons. The view of the classical tradition that measurements can be taken without any theory or without any preconceived notion of the world is entirely mistaken. Theory underlies the taking of any observation in science. Theory also dictates the nature of the measurement and a great deal of the way in which it is made. These relations between measurement and scientific theory will be illustrated in Case Study 11.1.

Case Study 11.1: River channel morphology

Regularities in the morphology of river channels and valleys have fascinated natural scientists for centuries. Their study provides an example of the history of the development of ideas and the relation between theory and observation. Nearly two hundred years ago John Playfair wrote:

> The structure of the valleys among mountains shows clearly to what cause their existence is to be ascribed. Here we have first a large valley, communicating directly with the plain, and winding between high ridges of mountains while the river in the bottom of it descends over a surface, remarkable, in such a scene, in its uniform declivity. Into this open a multitude of transverse valleys, intersecting the ridges on either side of the former, each bringing a contribution to the main stream proportioned to its magnitude; and except where a cataract now and then intervenes, all having that nice adjustment in their levels which is the more wonderful, the greater the irregularity of the surface. These secondary valleys have others of a smaller size opening into them and, among mountains of the first order, where all is laid out on the greatest scale these ramifications are continued to a fourth, and even a fifth, each diminishing in size as it increases in elevation, and as its supply of water is less. (quoted in Chorley et al. 1964, p. 62)

More recently this orderliness has been examined by a school of American workers which includes Leopold, Wolman and Miller (1964). One of the central problems with which these scientists have been concerned has been the relations between channel discharge, width, depth, slope and water velocity. They began to reason that the nature of river channels, their orderliness, could be explained by the mechanics of water flow and of sediment movement in channels of increasing size and changing discharge. The orderliness of channel morphology they expressed as a now famous set of graphical relations (Figure 11.1). Leopold and his co-workers suggest that the apparent simplicity and consistency of the relations in itself implies that some general (universal) explanation must lie behind the orderliness of these natural phenomena.

The consistency with which rivers of various sizes and in various physiographic settings make the adjustment to increasing discharge downstream suggests that

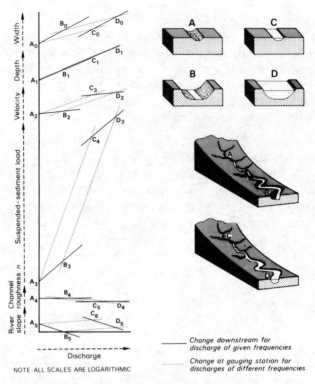

Figure 11.1 Relations of average river width, depth, velocity, suspended sediment
load, roughness and slope to discharge at a station and downstream
Source: after Leopold et al. 1964

there is a common general tendency or physical principle governing these adjustments. Because the channel adjustments are closely related to the profile of the
river, the latter must be considered before seeking a more general statement of the
physical principle governing the adjustments of both gradient and form. (Leopold
et al. 1964, p. 248)

A more recent examination of patterns of stream morphology by Klein (1976) has
revealed that the linear relations described by Leopold et al. are only simplifications
of compound relations. The width–area relation, for example, is different for different sizes of basins (see Figure 11.2). Klein speculates on the reasons for such differences in terms of the sources of channel-forming flows and the velocity with which
water moves by various pathways. Such an analysis follows not only from the work of
Leopold and Maddocks (1953) on channel morphology but from that of Hewlett and
Hibbert (1963, 1967), Whipkey (1965), Dunne and Black (1970a, 1970b), Freeze
(1972) and many others on the production of runoff. On channel morphology
Klein writes:

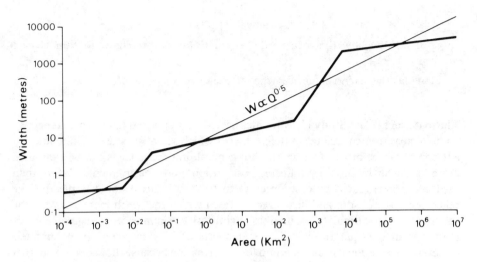

Figure 11.2 Channel–width–basin-area relation
Source: after Klein 1976

The break point of the line can be tentatively explained in terms of the contribution of runoff to the channel form: (a) the headwater overland flow; and (b) subsurface flow. Channel flow velocity can be regarded as constant relating to the
subsurface velocity and *vice versa*. As a result, the response of the basin to subsurface flow occurs several hours after the rain has started while the runoff from
headwater sources is already flowing down the channels. Assuming a constant
channel velocity of 1 m/sec and a delay to peak subsurface flow of 8 hrs, runoff
from headwater source areas will flow 28.8 km along the channels before the subsurface flow peak reaches the channel banks. The empirical relation between
mainstream length (km) and basin area (km^2) is: $L = A^{0.6}$, therefore if $L = 28.8$ km,
$A = 270$ km^2. (1976, p. 29)

It is at this basin area that Klein has found a break point in the channel-width–basin-
area relationship.
Consider the development of these studies on river channel morphology.

Which of the conclusions presented above are most easily tested?

Speculate on how the measurements relate to the problem situation of each
study.

How did the scientists decide what to measure?

The material in Case Study 11.1 provides us with an important historical perspective
on how measurement relates to the problems which scientists study. Three important points can be made. The first is that general, imprecise ideas can be examined
more rigorously by carefully collected observations and measurements. Rather than
say that rivers gradually become larger downstream we can say, for example, that the
relation between width and discharge is one in which the width increases with the
square root of the drainage basin area. Quantitative relations provide greater information than non-quantitative. The statements of Playfair about river channels
increasing in size downstream contain less information because they could mean that
any one of many different relationships could exist. Once expressed by Leopold,
Wolman and Miller as an equation, the relationship was more easily testable. Klein
subsequently produced a relationship which was even better testable.

The second major point which Case Study 11.1 illustrates is that scientists do not
work in isolation. The fact that both Hutton and Leopold found themselves inves-
tigating streams is not a haphazard occurrence. Hutton's observations of fluvially
eroded landscapes derived from his scepticism of the then current ideas of how
landforms were produced. He was propounding the idea that landscapes were
gradually eroded by running water and not shaped by catastrophic events. The
interest of Leopold and his co-workers stems from a complex situation which
involved the desire to explain landforms by a set of simple mechanical (Newtonian)
theories such as had been worked out by engineers to explain flow and sediment
transport in simple channels and pipes. They also believed, as did Hutton, that the
gradual wearing away of the land and the development of erosional and depositional
forms was susceptible to logical testable explanations. Their observations thus had
an ancestry in Hutton and Newton. There was also an ancestry in large-scale theories
of landscape evolution such as that of W.M. Davis who had provided a description of
landform development which was under attack. In other words the latter-day
geomorphologists also had an ancestry in ideas which they sought to criticize and
overthrow. Their measurements provided specific relations which tested old ideas. In
turn Klein's interest followed from a critical examination of Leopold's description of
downstream changes in river morphology. Many other studies had raised doubts
about the simple relations and Klein's contribution was to develop a general theory
for the systematic variation in channel morphology which had been observed.
Similar historical relations can be found in all science which illustrate how theories
and the growth of knowledge depend on what has gone before (see Chapter 8).

The third major point which case Study 11.1 illustrates is that scientists do not
make observations and decide later what they mean, nor do they merely make observations in the hope that one day they or someone else will develop a new idea or

theory from them. Nor when they make an observation do they do it in any arbitrary, uncontrolled way. Playfair, Leopold, Wolman, Miller and Klein were all driven by the need to explain something which previous theories did not explain. They knew what they were looking for and the observations they made related specifically to their problem.

As Case Study 11.1 shows, testing of theories provides the reasons for making observations but the relationship between observation and theory is deeper. Theory also lies behind the *way* in which observations are taken. Using the material of the case study, we must ask ourselves whether, for example, widths or areas were measured at any state of flow or at any position? Was any river used or was a choice made? In other words, according to what rules were observations taken? These rules are provided by the theory in question and its background of related theories.

In general, scientific theories will not be about relations between attributes but about relations between objects or relations between systems. The problem for the scientist is to translate his theory into specific relations between measured attributes. The jump from a theory to the things to measure is provided by the bridge principles of the theory (see 'Theories and hypotheses' – Chapter 1). It involves deducing from the theory which attribute can be measured in order to represent the conjectured behaviour of the object or system and how that measured attribute relates to other attributes. Statements about measured attributes and their relation then provide the basis for a critical experiment to test the theory. In order to illustrate this, consider the following imaginary situation.

Suppose the scientist held the theory that plant growth is controlled by the availability of light, such that increased light gives increased growth. Such a relationship holds because increased light is thought to stimulate photosynthetic activity (assuming that there are no limiting factors on growth). In testing this theory our first problem is to decide what to measure? Let us take for granted that we have a laboratory situation with reproducible conditions and a controlled light source, and that we are dealing with genetically identical plants. The availability of light can be measured in several ways such as in terms of light intensity, duration or periodicity. Plant growth can be measured as biomass (total weight), height, girth or leaf area. If we used the relation between height and light intensity we would probably find that there was a negative relation, since the response of the plant to reduced light could be to grow taller. If we took girth and light intensity then a positive relation might be found because in growing taller under conditions of reduced light the plant reduces the resources available for other growth. Using different measures of light availability might lead to yet other relations because of the adaptation of the plant to particular light conditions. What this shows is that a variety of relations between growth and light availability can be found by taking different measures of these two attributes. The implication is that the design and interpretation of experiments or observations can only be successful if we have deduced the consequences of our theories *in terms of the variables to measure*. The variables we choose to measure and the measurements we make have a direct relation to our theory. If we simply go ahead and make measurements we have not acknowledged this relation, but it exists nevertheless. It is easy to see from the example above how dangerous neglecting this relation can be in our assessment of theories, since the relation which is found between growth and light availability depends entirely on how these attributes are measured.

In summary, the measurements we make of objects or systems are made within the context of some theory. That context is provided by *deductions* from the theory. Measurements which fulfil the conditions required by the theory can be termed *valid* measurements. There are further problems arising from the manner in which measurements are taken, which are explored below. These relate to the theories of statistics. The rules these theories provide for making observations are contained within the principles of experimental design (Chapter 12).

2. Measurement and statistical theory
Any measurement is subject to error. There are many kinds of error and in most situations where measurements are taken the scientist has to take account of them and estimate their effect on his results. This can be done by reference to the theories of statistics. Even the simplest measuring exercise can be fraught with difficulties as the following case study shows.

Case study 11.2: The platean wall
In Thucydides' account of the Peloponnesian War in the fifth century BC he describes the siege of the Plateans and the ingenious plan to escape over the wall which had been built around their city:

> In the same winter the Plateans who were still being besieged by the Peloponnesians and the Boetians, finding themselves in distress as their provisions ran out, and seeing no hope of help coming to them from Athens or any chance of survival by any other means, made a plan with the Athenians who were besieged with them by which they were to leave the city and do their best to force their way over the enemy's surrounding wall . . . Their method was as follows: they constructed ladders to reach to the top of the enemy's wall and they did this by calculating the height of the wall from the number of the layers of bricks at a point which was facing their direction and had not been plastered. The layers were counted by a lot of people at the same time, and though some were likely to get the figure wrong, the majority would get it right, especially as they counted the layers frequently and were not so far away from the wall that they could not see it well enough for their purpose. Thus, guessing what the thickness of a single brick was, they calculated how long their ladders would have to be. (Thucydides, see Warner 1954, p. 172)

Over two hundred men made their escape and returned to Athens. Eventually the city surrendered, however, and the Spartans mercilessly put to death no less than two hundred of the Plateans and razed the city to the ground.

Why do you think a lot of people were used to count the bricks?

Why do you think they all counted the bricks many times?

Do you agree with Thucydides that the majority would get the figure right?

Even with a very large number of repeated estimates of the number of bricks, why might the answer still be wrong?

Is Thucydides' account altogether consistent? Do you think he has told us the full story correctly?

Using the case study above, we can illustrate some of the types of error which arise in measuring. The simplest kind of error is that involved in measuring an object when it is unaffected by the measuring process, as occurs, for example, in counting the number of bricks in a wall. If we repeatedly measured the bricks in the same wall we would produce a sequence of values which would not be identical. This happens because no sequence of operations can be reproduced exactly. Such errors have no effect if the level of precision that we use is coarse. If there are 20 bricks in the wall, then it is unlikely that a competent, sane worker will ever measure its height as anything but 20 bricks. However, if the blocks in the Great Wall of China were being counted, it is very unlikely that the same result would be obtained twice, and even to the nearest 1000 blocks some variation would occur. This type of variability is called *pure measurement error*.

If the bricks were counted at the same point under the same conditions by each member of a group, it is likely that they would produce different results. In this case the error arises in two ways. In addition to measurement error there will be an error due to the differences between the people counting. This is called *operator error*.

In contrast to the situation described above, we can consider the case of an object which, even if the measurement process were perfect, cannot be measured repeatedly either because the measuring process destroys the object or because only part of the object can be used and repeated measurements are of different parts. If, for example, we measure the weight of a plant, roots and all, then we must take the plant from the soil and thereby destroy it or alter its growth so drastically as to make any further measurements meaningless. A more common situation is one where only a part of the object of interest can be measured, as in the case of measuring the size of sediment in a river channel. In these situations we *sample* from a *population*. In the case of a river it is impossible, not to mention undesirable, to take all the sediment. Thus only a small, convenient sample is taken. If we want to repeat the measurement of sediment size we take further small amounts. Such measurements would vary owing to the differences between the samples, even though the same sediment is being sampled. The error these differences cause in estimating the sediment size of the entire deposit is called *pure sampling error*.

Normally we have no knowledge of the sediment size in an entire deposit or the number of bricks in a wall, so that the magnitude of the sampling error remains unknown. If we were being strict we could say that taking a sample is inadequate since we do not know that the sample is like all other possible samples or like the population. We can nevertheless approach this sort of problem by dealing instead with probabilities. This approach to the problem of sampling error is through the methods of *statistics*. The theories of statistics provide the assumptions about how samples relate to populations and thus estimates of the sampling error in measurements. This is the basis for deciding whether two different measurements represent different populations or are different because of sampling error. Such reasoning about the separation of real differences from error permeates all empirical science. What statistical theory allows us to say in absolute terms is the probability that differences between measurements are due to error. If that probability is sufficiently small then we would conclude that they are not due to error but to the differences in the populations from which they were taken.

Let us consider an example of the kind of problem which may be tackled using statistics. It is assumed that the reader has a knowledge of elementary statistics. Say we hold the scientific theory that the angle of slope of a river channel depends on the size of particles in which that channel is eroded. In testing this theory it is necessary to assume that we can choose comparable sections of channel in which we can take measurements. This is because channel slope can depend on other factors and we wish to eliminate their effects. We could test the theory by examining whether channels which in all other respects were the same, had a steeper slope in larger sized material.

We might, in the first instance, choose two channels with different angles of slope. The slopes should be sufficiently different to show an effect if our theory were true but remain within the range of channel slopes where flow conditions were comparable. If all the clasts in a stream were of the same size then a sample of one clast from each stream would be sufficient to answer the question. The only kind of error that might arise would be measurement error. In reality this situation is unlikely. Normally there is a variety of clast sizes in any one stream and we do not know what the pattern of variation is. Therefore to achieve some representation of the variability we take a larger sample, let us say 50 clasts. When we have taken two samples of 50 clasts, one from each of the two streams, they might be all of different sizes but it is possible that every clast in the first sample is larger than every clast in the second. Then it would seem that the size of clasts in the first stream is larger than in the second. It is possible, though very unlikely, that we picked only the very largest clasts in the first stream and only the very smallest in the second and in reality there is no difference between the streams, or the difference is not what our samples show. We have no way of knowing, however, that such a situation exists and we have to make a decision on the basis of the results we have found.

By careful choice of site and samples we may eliminate the effects of obvious major influences on clast size. Once such experimental controls have been established, we can consider the remaining influences, which are generally unknown, to have random effects on the size of clasts, that is, that there are no systematic effects which will be revealed by a sampling scheme. There are two consequences of this situation. First, the sampling errors under random influences have a frequency distribution which is bell-shaped, described as the normal curve of errors. Secondly, if we are to take unbiased samples, then we must take them randomly. Such samples are those whose selection procedure contains no systematic elements and which therefore cannot in the long term lead to unrepresentative measures of populations. A single random sample may be most unrepresentative of a population but the chance of this can be calculated if there is no systematic element in the sampling procedure. These two conclusions from statistical theory are central to experimental design. If we assume that the unknown populations of clast sizes have a specific distribution and that random samples provide unbiased estimates of their properties, then we can take samples and yet make statements about populations. This leap is called *statistical inference*.

Going back to our samples of 50 clasts, let us assume that they are taken randomly from normally distributed populations. If the two samples are different we still cannot say definitely whether they come from populations which are different but we can give the chance that they do. How can this chance be calculated? Let us assume in the first place that there is no difference between the populations. In statistical theory this assumption is called the *null hypothesis*. We have to begin with such a hypothesis because we must not assume that the samples are different. In other words we must not assume a theory being true as the basis for a decision about whether it is true or not. We assume instead that it is false. Thus we proceed by assuming no difference between sample means and then look for any evidence which would allow us to refute this idea. We assume that there is no effect of slope on clast size and then calculate the chance of taking samples with differences in mean clast size as great or greater than we actually found. If the chance of finding a difference as large as we do is sufficiently small, then we would be prepared to accept the hypothesis that there is a real difference between the streams. On the other hand, if the chance of finding a difference as large as we have is high, then it would be more rational to accept the conclusion that the difference between samples is due to sampling error and has nothing to do with the streams themselves. Rather than say, 'We have got a difference between samples which is very unlikely (given that the populations are not different)', we say instead, 'The difference we have is so unlikely that we reject the assumption on which it is made'.

Two points worth noting in this procedure are, first, that in testing a theory we assume that it is not true and only accept the theory if this assumption is untenable on the basis of evidence, and secondly, that the theory is not tested in isolation. We only accept the theory if we can find evidence that would refute its competitors (Strong 1982). Logically such a method follows the methodology of the science described by critical rationalists. Since we cannot verify anything, but can only refute it, we should be critical, and minimize the chance of accepting a theory as true when it is false. We aim to refute our theory by setting it against a competing theory, and only accept the original theory if on the basis of the evidence the competing one, the null hypothesis, is improbable (see 'Multiple working hypotheses' – Chapter 3).

The theory we are dealing with relates slope of channel to clast size. This relation has been expressed verbally. But in the description of our observations the relation was expressed quantitatively as a *difference* between sample *means* of clast sizes from channels of two different slopes. Slopes and clast size can both be measured on ratio (R) or ordinal (O) scale. Since there are various combinations of scale which are possible, the relation can be expressed quantitatively in a number of ways other than as a difference of means. In other words there are several statistical models which can be used to decribe or test it. We could do any of the following:

1. perform a t-test of difference of mean clast size (R) at two sites with different slope classes (O, R);
2. perform an analysis of variance of clast size (R) at several different slope classes (O, R);
3. find Pearson's correlation coefficient between mean clast size (R) and slope (R) at different sites;
4. find Spearman's correlation coefficient between modal clast size (O) and slope (O);
5. perform a Kolmogorov-Smirnoff test on the frequency of clast size (R) at different slopes (R, O).

Thus the relationship can be explored quantitatively in many different ways using various statistical models but each model has its own assumptions about the data and the populations from which samples were taken. These assumptions relate to the number scale, the normality of populations and the size of sample. Therefore the scientist must choose the method of statistical analysis *before* taking measurements. Only when measurement fulfils the requirements of statistical theory is it said to be *adequate*.

Try Exercise 11.2.

Conclusion

There used to be a view that the physical geographer is educated 'through the soles of his boots'. The modern-day equivalent might be that education is accomplished through the learning of techniques of measurement and analysis. This is carried to extremes in the preoccupation with quantification and statistical methods. The acquisition of such skills becomes somewhat disappointing when it is associated with the belief that measurement and statistical analysis are the hallmark of the transition of physical geography to a science, and when the techniques are used in a way which pays no more than lip service to scientific theories. In fundamental respects early workers such as Du Buat, Hutton, Playfair, Darwin, Wegener, Gilbert and Davis understood the methods of science better than their descendants, since they never regarded measurement as an end in itself or as being divorced from theory. The error in much contemporary work is the failure to recognize that measurement and statistical analysis are tools by which theories are tested critically. To try to divorce measurement from theory is to fail to understand the theory content of measurement and the measuring process. To use statistical methods in the attempt to let the data speak for themselves is to misrepresent the logical bases of statistical methods and hence of scientific method. Quantification and the use of statistical models must be seen as extensions of scientific method.

Within the context of critical science the measuring process can be seen to hold a complex relation to scientific theories, statistical theories and testing. When measurements are made it is important to be aware of these relations and of the limitations which each type of measurement possesses. Number scales and processes of measuring lay constraints on the relations which can be established and thus on the extent to which empirical evidence can be used to make judgements about theories. Thus in deciding what to measure and how to measure there must be an adjustment between the demands of a test and what can be learned empirically.

CHAPTER 12:
EXPERIMENTAL DESIGN

'What am I going to learn from this?'

Introduction

Scientists take measurements in response to some problem of theory testing. In approaching such a problem it is necessary to exercise great care in the taking of measurements in order to avoid the errors and ambiguities inherent not only in the measuring process but in all aspects of testing. This is achieved in the controlled and coordinated process of experimentation. By way of introduction let us consider three problems and what is involved in answering them.

1. Is Mozart a better composer than Walter Piston?
2. Is Mount Everest the highest mountain in the world?
3. Are erosion rates greater at higher altitudes?

The answer to the first question is largely a matter of opinion. If there is an answer it is not one which could be reached through an experiment. We cannot ask of Mozart or of Piston that they produce pieces of music to compare even if we were confident that we could judge their quality unambiguously. In contrast, there is something we can do to answer the second and third questions. We can make measurements. However, these two questions differ in their demands. In answering the middle question, once having defined what is the highest mountain and agreed on a satisfactory way of measuring it, then it is a simple problem to find the answer. There are of course uncertainties in measurement and in definition, but within these constraints there are no real problems. In the final question, however, there is great uncertainty, arising from the relation between erosion rates and altitude, about exactly what erosion rates are and what factors control them. The measurement of erosion rates is fraught with difficulties because of problems with techniques of measurement and with the variety of conditions under which erosion takes place. Thus, answering the third question requires a great deal more care and thought.

Science is generally concerned mainly with the third type of question and an important part of scientific methodology relates to the way we go about answering such questions. This methodology involves argument and arguments are based on logic. The procedure for taking measurements in uncertain situations in order to test ideas in such a way that we can confidently enter into arguments about them is called *experimental design*. This chapter deals with the design of experiments. There can be considered to be three stages in the design and in the execution of experiments: the statement of the problem, controlled measuring and interpretation. First, however, it is necessary to consider the different ways in which experiments can be conducted.

Types of experiment

Scientists design experiments in different ways with different attitudes of mind. Medawar (1979) has attempted to summarize the different kinds of experiment which can be undertaken. In this book we have placed great emphasis on adopting a critical attitude to theory. Medawar described experiments which are designed to test theories critically as Galileian experiments, after Galileo, who reputedly performed a critical experiment on gravitational acceleration by dropping balls of different weight from the leaning tower of Pisa. Since the essential property of scientific hypotheses or theories is that they exclude or prohibit something, a proper scientific test exposes the theory to a situation in which whatever is prohibited will arise if it can. If it does arise then the theory can be refuted. This is the essence of the Galileian experiment.

Sometimes, however, the same situation may be achieved by a thought experiment. Thought experiments can be as valuable to the scientist as those which involve physical objects. Medawar describes such experiments as Kantian experiments. These are characterized by the preamble 'Let's see what would happen if we took a somewhat different view.' These experiments are mental or computer-simulated explorations of the consequences of ideas. They often involve deducing the consequences of such things as changing the initial conditions of an explanatory hypothesis or working out the fuller implications of a theory. Medawar contrasts this kind of experiment, which conforms to the critical rationalist approach to science, with those based on other attitudes.

The approach to experimentation based on the classical tradition is one which stands out in particular contrast to the Galileian experiment. Medawar describes experiments based on the ideas of verification and induction as Aristotleian and Baconian respectively. Aristotleian experiments are undertaken to demonstrate the truth of ideas or theories. Situations are created or observations selected which provide results which fit the preconceived idea, and which are taken to demonstrate that it is true. Baconian experiments are responses to the question, 'What will happen if . . .?' In other words, by observing the world in different ways, or observing what happens if it is stimulated in some way, we can accumulate different instances of the world's behaviour. Such observations will then add to the body of factual information and, according to the inductivist belief, so our understanding will grow. These experiments are not critical, however, because they are based on the idea that the truth about the world is there to be observed and that to establish our theories we have to produce circumstances or make observations which verify them.

In order to become familiar with these types of experiment consider the examples in the following case study.

Case Study 12.1: Types of experiment
Consider the four examples of experiments and identify which type of experiment each one represents. For each one say which particular characteristics you used in order to distinguish it.

Experiment 1   Local variation in background water quality

A study was undertaken by Webb and Walling of the chemistry of small streams in part of Devon, England. The purpose of the study was to evaluate the extent and controls of local variations in stream water quality. The study area, the catchment of the River Exe, contained a variety of geology, land use and relief, which were considered to make it worth while. Quality was measured using specific conductance (μmhos/cm at $25\,^\circ$C). Samples were collected during a period of dry weather so that conductivity levels were relatively constant. Over 500 sites were sampled, which provided a very thorough coverage of the study area. Webb and Walling wrote:

> The study demonstrated a marked spatial variation in background water quality within the Exe catchment. A broad spatial pattern was also apparent ... [with high values] ... exhibited by streams draining the south east portion of the basin and low values are associated with upland areas of (the north).
>
> The spatial contrasts in water quality merit some attempt at tentative explanation. (1974, pp. 145–146)

Webb and Walling then presented mean conductance values for streams draining each of the major geological divisions and each of the main land-use types on the geological divisions. They showed that many of these values are statistically significantly different from each other. They continued:

> The above considerations clearly demonstrate that background water quality ... varies within the Exe catchment according to geology. There are, however, several difficulties involved in any attempt to ascertain the precise *extent* to which water quality is influenced by rock type.
>
> Land use would seem to influence the spatial variation of specific conductance levels encountered in this study although because geology and land use are closely interrelated it is difficult to isolate individual effects. The trends discussed above can be tentatively accounted for in terms of fertiliser application, animal grazing and the degree of soil disturbance but more detailed investigation of nutrient cycling would be necessary to produce firm conclusions.
>
> Interdependence of environmental factors is even more evident when one attempts to consider the effects of topography. Topography will be closely related to both geology and land use and although it could be argued that the steep relief of Exmoor would condition low solute concentrations, because of reduced flow through times, the influence of lithology would seem more important. The effects

of topography are probably to accentuate the influence of rock type and to condition local small-scale variations. (pp. 147–149)

Experiment 2   Planation surfaces of North Cardiganshire, Wales

As we have seen in Case Study 4.2, until the 1960s British geomorphology was dominated by the study of denudation chronology. One of the dominant ideas of this approach was that of the existence of a series of planation surfaces in upland Britain. The underlying ideas of the study of planation surfaces is the concept of the cycle of erosion, championed by W.M. Davis. During the 1960s and 70s with the spread of quantitative methods there were several attempts to identify and describe these planation surfaces by statistical techniques. One study by Rodda of Welsh surfaces began by recognizing that:

> ... different workers have distinguished a variety of plateau surfaces usually at
> different levels and even the idea of a single concave surface has been put forward
> ... By distinguishing three surfaces E.H. Brown has advanced what is perhaps the
> most rational solution ... There remains some doubt about the validity of this
> three-fold division. Trend surface analysis offers a convenient and objective
> method for examining the landscape and determining the number of surfaces that
> comprise it. (1970, p. 107)

Trend surface analysis is a form of three-dimensional regression analysis. It was applied to data on all summit heights and heights of closed contours in an area covered by the 1:63,360 Ordnance Survey Sheet 127 (Aberystwyth). Trend surfaces were computed and the goodness of fit of the surfaces and the patterns of residuals used to judge if there was one or several surfaces and at what heights they were found. Linear, quadratic and cubic surfaces were computed. Residuals from the trend-surfaces, differences between the actual height and the height of the trend-surface, showed common patterns for the three computations (see Figure 12.1). Rodda's conclusions were:

> [The] major features ... consist of a series of north–south strips alternating in the
> sign of the residual ... The existence and repetition of this systematic pattern in
> the distribution of residuals indicated that some feature or features in the
> landscape were not being taken into account by the single surfaces. Indeed the
> sequence of positive and negative values from west to east could best be explained
> by a series of steps ... Two such changes are apparent ... Of course it could be
> argued that the pattern of residuals resulted from the north–south alignment of
> some of the river valleys ... but the effects of these valleys must have been
> marginal.
> To investigate further the possibility of the landscape being composed of three

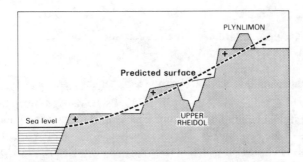

Figure 12.1 Planation surfaces and their relation to the predicted surface
           Source: after Rodda 1970

planation surfaces, the original data were divided into three so that three separate trend-surface analyses could be performed.

From the results of these analyses, it is evident that all three trend surfaces are separately less significant ... than the single one ... On the other hand, .. the accountable variance is higher.

Thus it is evident that Brown's three-fold division of this part of north Cardiganshire is justified. This analysis substantiates his results, although it also indicates that a single concave surface is almost as appropriate. (pp. 109–111)

Experiment 3   The generation of surface runoff

During the 1950s and 60s several field studies of the runoff process in humid temperate areas had shown that the generation of runoff by overland flow did not occur. Several mechanisms had been put forward to explain runoff and the principal ones were:

– overland flow due to surface saturation from below
– subsurface storm flow
– overland flow from near channel partial areas due to surface saturation from below.

The paper by Freeze (1972) provides a theoretical analysis to complement the field studies. In particular it presents the results of mathematical simulations to help identify the conditions giving rise to those mechanisms that are supported by the field evidence. The paper concludes:

Considerable evidence to support the claim that horizons of shallow surface soil of high permeability are a common occurrence in both forested and agricultural watersheds exists but such soil conditions do not guarantee the eventuality of subsurface storm flow. Theoretical simulations of runoff generation in upstream source areas show that there are stringent limitations on the occurrence of sub-surface storm flow as a quantitatively significant runoff generating mechanism. The occurrence of subsurface stormflow is feasible only where convex hillslopes feed steeply incised channels, and even in such instances a threshold saturated hydraulic conductivity exists below which subsurface storm flow cannot be important.

On convex slopes with lower permeabilities, and on all concave slopes, direct runoff through very short overland flow paths from precipitation on transient near-channel wetlands dominates the hydrograph. In these expanding wetlands surface saturation occurs from below because of vertical infiltration toward very shallow water tables rather than by downslope subsurface feeding. (p. 1282)

Experiment 4   Succession in British lakes

Succession is the idea of an orderly change in vegetation in response to changes in the environment brought about by the plants themselves. It is supposed to be progressive and predictable. A common example of a succession site is a small lake, whose pattern of change is caused by the advance of vegetation from the edge in a successional man-ner. Walker's (1970) investigation of British lakes is reported by Colinvaux (1973) as follows:

The record of infilling of an old pond should be preserved in its mud . . . It should be possible to trench or bore into the mud beside an old pond and reconstruct the sequence of the plant communities that occupied the site . . . Walker collected together (such) records of borings at the edge of lakes and ponds from various parts of Britain and has used them to test the hypothesis that infilling of ponds results from the action of plant communities appearing in a predictable orderly sequence . . . A first part of his finding was that the fill in most of his ponds was material brought from outside by erosion and runoff and was not autochthonous organic matter deposited by the plant communities themselves. The plants had passively responded to the infilling; they had not caused it. And his second finding was that there was no predictable sequence in which the plant communities appeared.

Walker came to another provocative conclusion; that the infilling of British ponds resulted not in the establishment of the local formation [i.e. plant com-munity], but in peat bogs. (pp. 89–90)

The value of an experiment is not to be judged by the quantity of the data generated or the novelty of the techniques used, nor is it to be judged by the claims of the scien-tist alone. The main guide in judgement is whether or not it advances knowledge. This rests on the extent to which an experiment provides a critical test of a theory. In making a judgement the objectives of a piece of research must be considered and compared with its concrete results. In the first example about water quality Webb and Walling's objective is merely to find out something about stream water chemistry without any mention of theory, which fits the description of a Baconian experiment. The study did however provide some information on the spatial pattern of stream water chemistry which excluded many possible relations. The initial statement of purpose follows the classical approach of looking for regularities in the hope that ideas will emerge and in this case none emerge. The concluding remarks consist largely of speculations on the possible relations of certain factors to stream water chemistry which could have been made without any measurements. Indeed the conclusions should have formed the *basis* for an experiment to test opposing theories.

Rodda's statistical analysis of planation surfaces, in contrast, does reach specific conclusions about Brown's threefold division of Welsh surfaces with the rider that a single surface may be as appropriate. The point here is that Rodda only looked for surfaces. This is clearly inconsistent with the supposed 'objective' method of trend surface analysis, which by its nature can produce only surfaces (Tarrant, 1970). It can be the basis of rejecting the idea of surfaces altogether, only if there is some criterion of goodness of fit which distinguishes surfaces from features which are not. Thus Rodda's study can only support or verify the existence of surfaces and is an example of an Aristotleian experiment.

Freeze's analysis is different from the other three in that it is almost entirely accomplished on a computer. It is a nice example of a Kantian experiment exploring the consequences of various theories in order to test them. The method provides discriminating arguments about the validity of apparently opposing theories and reaches specific conclusions to the specific problem Freeze posed. Similarly Walker set out to test the idea of succession with a specific proposal. As with Freeze's computer simulation, there could have been various outcomes including the possibility of that outcome in which a theory would be judged to have failed, and in each case a theory did fail. Finally, one of the most impressive contrasts between the first two examples and the last two is in the fruitfulness of their consequences. The first two studies go no further than the problems they posed whereas the last two generate questions beyond those which gave rise to them.

Experiments provide the basis on which we can argue and come to decisions about a theory. They involve consideration of the implications of a theory through its bridge principles, the process of measuring and the interpretation of observations. All of these need to be part of a coordinated plan. The scientific theory, the statistical theory underlying the observations and the logic of argument demand that observations be made according to certain rules. The need for coordination and compliance with rules demands control in the taking of observations. To achieve this control there has to be some prior, conscious selection of actions following a set of rational procedures. There has, in other words, to be *design* in our experiments. This design has three closely related stages by which control is achieved. These are: the statement of the problem, measuring and interpretation. In order to explain some of the methodological problems of experimentation in physical geography, and to identify some of the links between practice and the philosophical issues which have gone before, each stage will be considered in more detail (see also Cox 1958; Fisher 1966; Hicks 1973; Lacy 1953; Jeffers 1978; Yates 1970).

Statement of the problem
In designing an experiment, before any measurement can be undertaken the experimental problem needs to be specified. There are four aspects to this considered here. The first is the problem of deciding what to measure and why.

1. The response variable
The *response or dependent variable* is the measured property of the objects or situations which are the focus of interest. Its definition follows from the scientific theory and the bridge principles which the theory contains. A theory may be about a particular class of thing or some attribute. In general the thing or its attribute may not be measurable. It is then necessary to deduce from the theory, by way of its bridge principles, what measurable properties can be used in order to answer questions about it. There may be several steps in this process and at each one deductive logic dictates how a surrogate variable relates to the property expressed in the theory. For example, the scientist may have a theory about successional stages in ecosystem change. It is not possible to measure such a stage and so plant communities are defined as the vegetational expression of the changes taking place. A community also cannot be measured, it is a concept like the 'stage' in geomorphology; but plants can be measured and their frequency or cover can be used to define communities and thereby the stage in succession. Thus there are a number of steps the scientist might go through in deciding from a theory what it is he wants to measure. The measurement is deduced from the causal relations which the theory contains.

The next problem is, on the basis of the theory in question, to define the population to be sampled. The population is that about which the scientist is seeking to make statements on the basis of the experiment. It is the class of things referred to by the theory in question. It could be, for example, all rivers, or rivers in semi-arid areas, or small, ephemeral channels in semi-arid areas. Whatever the case, for any class of objects the response variable should be measurable. The scientist can speculate on erosion rates in the Quaternary but are they in fact measurable? Similarly if a relation such as one between channel slope and bed material size is to be examined it is important that the range of material sizes and slopes to which the theory relates actually exist and are available for measurement.

A problem that often arises in physical geography is the temptation to use a response variable because it is easy to measure or has been used frequently before. In some areas of the subject there are response variables which are traditional: slope angle, particle size distribution, average wind speed, and plant species lists, for example, which are used frequently with no real regard for their relation to any theory (see

Chapter 11, especially the conclusion). The problem which the experimenter has is to deduce the relations that exist between the response variable and the thing of interest if the theory were true.

2. Factors to vary

Any theory deals with how one thing affects or causes another. The response variable is derived from the theory and so are the variables which represent the things affecting it. These variables are usually termed *independent variables*. If we are considering, for example, how a clast changes in size as it travels along the bed of a river, then intuitively we can recognize that the changes in size it experiences can be caused and affected by many factors. These are the independent factors controlling clast size. How are they represented in an experiment? Once again it is the bridge principles of the theory under test which specify how the independent variables are determined so that they represent the independent factors. Like the response variable the independent variables must be measurable and they must be represented in the available experimental situation. When one factor is being dealt with, experimentation demands the control of other factors. How control of these extraneous variables is achieved is described in the section on 'Controlled measuring' below. At this stage it is sufficient to note that all the problems which relate to the specification and measurement of the response variable also apply to the independent variables. However, there are two additional problems. First, if the dependent variable is responding to the independent variable, then values of the independent variable must be selected within a range which will produce a measurable effect in the response variable. Secondly, there is the problem of *interaction* between independent variables. This is the effect of one independent variable on another. The effect of one independent variable considered alone may include part of the effect of another and thus make it impossible to decide the extent to which it affects the response variable. If a theory is about two simple factor effects, then the experimental situation should allow the separation of interaction effects from direct effects. What this means is that when multiple factor experiments are undertaken, the number of measurements required for full analysis of effects increases rapidly.

In addition to the problems outlined above, there is the problem that natural systems depend on specific sequences of events for their characteristics. They have a history which is significant but often poorly understood. This makes it difficult either to use historical factors as experimental factors or to control them as extraneous ones. For this reason, or for any other, if we cannot perform a satisfactory experiment because of problems with response variables or factors, then we must design a new one. Rarely can a scientific problem or theory be tackled in one way only. Using imagination and ingenuity to design new experiments where old ones fail is one of the most absorbing roles of the scientist. It can involve developing new techniques of measurement, inventing new response variables, exploring further the logical consequences of theories or devising new analytical models.

3. Statistical models

The relation between experimental design and the statistical model is two-way. The choice of statistical model is constrained by the data which can be collected but the model, once chosen, dictates the requirements of the data collection. A major part of experimental design is the mating of data to statistical model. This is why a consideration of the statistical model must enter at an early stage in design. However, there is nothing sacrosanct about any particular statistical model, as we saw in Chapter 11. It is useful to realize that many statistical models are closely related. Analysis of variance and regression models are equivalent ways of handling discrete and continuous data. The z-test, t-test and chi-square test are all limiting cases of the F-test (Mather 1964, pp. 46–49). Many non-parametric tests are based on the same idea of a uniformly distributed random variable in the null hypothesis, and represent different devices for comparison with this distribution (Siegel 1956).

4. Anticipation

One vital aspect of design is to consider in general terms the possible outcomes of an experiment and to decide in advance, or to anticipate, what their consequences will be for a theory. For example, in relating stream velocity to distance downstream a negative relation might be expected, that is, that stream velocity decreases downstream. Such a relation could be based on the fact that channel slope decreased downstream. Since velocity depends directly on slope, in accordance with Newtonian principles expressed through theories of fluid dynamics, stream velocity should decrease with lowering angle. This would produce the negative relation with distance. However, if several streams were measured and a positive relation found, then how would this outcome be interpreted? Similarly, what with a constant velocity or a U-shaped or ogee-shaped relation? If the response to any of the expected relations is to retain a theory, then the problem arises as to what sort of result would be taken as the basis for rejection. If the outcomes of an experiment are never to be taken as a basis for rejecting a theory, then the experiment is pointless. Indeed, it is worse than pointless since it may provide false support to ideas.

An essential part of the critical rationalist view of science is that each experiment can throw up new unexpected problems, new facts and new refutations that demand a new explanation (see Chapter 3). However, the recognition of the new situation demands anticipation. The observations of Webb and Walling in Case Study 12.1 illustrate some of the problems arising from having little prior consideration of the outcome of an experiment. Webb and Walling stated nothing in their conclusions that they did not already know. The tentative explanations they provide are the tentative explanations they began with, which is shown by their justification for the study basin and by the statement of their purpose. It is not possible to justify such studies by claiming that they add to the body of knowledge by showing repeated instances of events. Repetition is the means of ensuring reproducibility, it is a test of some aspect of *performing the experiment*. It can provide a further test situation which may differ from previous ones, but it cannot in itself provide an argument for the validity of an idea or theory in the same way envisaged by the classical tradition (see Chapter 2).

To anticipate the outcome of an experiment does not mean that nothing new will have been learned. On the contrary, there is a great difference between speculating on possible results and having to explain those actually found, because the undertaking of an experiment will have raised new problems. The point about anticipation is to prevent the confirmatory experiment and to be prepared to meet the new situation. A nice example of a study which, with hindsight, we can easily see to have failed in this respect is Sandford's (1929) explanation of the mode of origin of the erratic rocks of the Oxford region (Figure 12.2). Ice sheets were supposed to have reached only as far south as the general area of Oxford but erratic rocks had been found from the south and west of Britain and there was other evidence of rocks from these regions along the southern coast of England. Clearly ice transport could not account for these erratics if the theory about the extent of the ice was true. Sandford's explanation was for transport by floating shore ice in some cases by the most unlikely routes and with high sea levels during periods of glacial advance. Rather than change his preconceived ideas Sandford invented the most improbable combination of ad hoc hypotheses to explain the distribution of erratics.

Sometimes scientists perform experiments hoping to confirm their theories. They might say honestly, if not truthfully, that they could produce results which refute a theory. However, if they have not critically tested the theory by allowing that situation to arise in the experiment which refutes the theory, then they have not performed a *bona-fide* experiment. The scientist must always ask the cynical question, 'What am I going to *learn* from this?'. If the answer is 'nothing new', then the experiment should be redesigned.

Controlled measuring

The second stage of experimental design is for the taking of measurements of the response variable and the variables which represent the factors which affect it. Since the scientific theory under test dictates that only certain factors are investigated, one of the main purposes of experimental design is to control for the effects of extraneous factors. There are four main methods of control (Lacey 1953):

   elimination
   equalization
   balancing
   randomization

Of these randomization is especially important and will be treated separately.

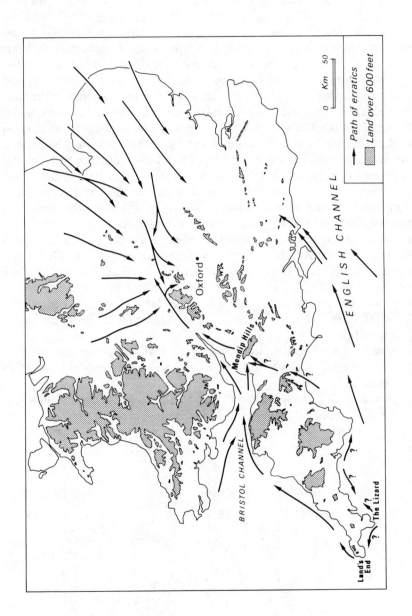

Figure 12.2 Glacial erratics of the Oxford region
Source: after Sandford 1929

Elimination, equalization and balancing

In order to illustrate these, consider a test of the idea that the size of gravel in a river decreases exponentially with distance transported. We might approach this, as others have done, by sampling river gravels from recognizable outcrops at different distances downstream from the outcrop. Such an approach introduces some control since it is possible to specify with some degree of accuracy the distance which a clast has travelled. However there are problems with taking one stream with one outcrop and plotting size against distance since not all extraneous factors are thereby controlled. For example, the eroding process may depend on the position along the river in which erosion takes place. What this means is that a pebble placed in an upstream position and travelling 1 km would be eroded to a different degree than the same pebble placed in a downstream position and travelling 1 km. The effects of distance travelled and position are said to be *confounded*. In other words they cannot be separated. This undesired effect cannot be overcome by eliminating the effect of position since wherever a pebble is in a stream it is in some position for which erosion differs from some other position. The effect of position can be controlled by *balancing*, that is, by placing a sample of pebbles in an upstream position and some in a downstream position, allowing them to move a fixed distance and changing the position of half of the pebbles in each sample to its original position and half to the original position of the other sample (Figure 12.3). This is necessary since the size of the pebbles is changing as well as their position and a comparison is needed between the swopped pebbles and a control group which was changing size over the same upstream or downstream section.

If we consider this experiment involving pebbles further, another recognizable, extraneous factor which may affect degree of erosion is the presence of a lake. This effect is controlled by *eliminating* it.

Other important recognizable extraneous factors on the erosion of clasts are lithology, shape and initial weight. All these factors can be controlled by *equalization*, that is, only selecting pebbles of the same rock type, weight and degree of rounding.

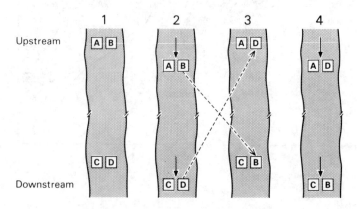

Figure 12.3 Abrasion of pebbles: balancing of effects

Randomization

If we consider the case of the pebbles more closely, then it is clear that erosion is a function of many other unknown factors including the random accumulation of impacts and pressures which result in one clast having different erosion from another. No direct control can be exerted on these factors since we are ignorant of them. The only method of ensuring these extraneous factors are controlled is to take measurements which will show no bias towards any external factor, that is, randomly. To sample randomly excludes the systematic effect of any factor and thereby controls them. Randomness is an assumption of all statistical models and is an essential requirement of sampling.

In experimentation a great deal of emphasis is placed on randomization. When known extraneous factors have been controlled then the statistical model assumes the complete randomization of the experiment. It is important to realize that randomization is part of the design. Simply using a table of random numbers at some stage in the sampling, or taking readings in a way which does not appear to be biased, is not a substitute for adequate design.

As an example of the problem consider an experiment on the effect of lithology on the shearing strength of soils. Consider four rock types: Granite A, Basalt B, Limestone C, and Sandstone D. An experiment requires a measure of errors and therefore replications are included. The variation due to replication gives a measure of the accumulated error from various sources. Since different rock types have slopes with dominant angles and slope may have an effect on strength, purely random sampling of slopes on, say, a spatial basis might result in variation between rock type being caused by variation in slope angle. Therefore a range of slope angles should be chosen from each rock type, say, flat I, gentle II, moderate III, and steep IV. We could measure the strengths of soils on rock types and slope angles assigned to each other as:

	Slopes		
I	II	III	IV
A	B	C	D
A	B	C	D
A	B	C	D
A	B	C	D

This layout is clearly invalid since the effect of rock type on strength cannot be distinguished from the effect of slope angle. Differences between A and B cannot be distinguished from differences between I and II. This is a completely confounded design.

A more satisfactory design might be to assign the rock group to the slope groups at random. This might result in:

Slopes

I	II	III	IV
C	A	D	A
A	A	C	D
D	B	B	B
D	B	C	C

This design suffers from the fact that strength is never measured for rock type A on slope III or B on I. Also any variation within rock type A may reflect variations in slopes I, II and IV. In general, since slopes are allocated to rock types differently, the random error (within rock types) may include some variation due to differences in slope angle.

A design in which each slope is used on each rock type only once and yet is chosen randomly is a randomized complete block design:

Slopes

I	II	III	IV
B	D	A	C
C	C	B	D
A	B	D	B
D	A	C	A

This design allows us to assess the variation in strength due to rock type and the variation due to slope as well as that due to experimental error, which the previous allocations did not.

Randomization should also be employed in the *order* of experimentation. If we measure the strength of soil in a shear box, there may be systematic errors in the apparatus, a change of operator or whatever. Therefore the order of measurement of the individual samples should be selected randomly.

Each experimental situation requires careful thought as to what extraneous effects might occur and the design of a sampling scheme to control them. Having decided that particular extraneous factors should be controlled by elimination, equalization or balancing, and that other extraneous factors are controlled by randomization, a choice must be made of the sites for measurements. The principal guide is the definition of a discrete unit of the object of interest. The definition of a discrete unit could be that of a sampling station in a river. In other studies it could be a soil unit, a slope facet, a river cross-section and so on. You may want to try Exercise 12.1 at this stage in order to consolidate your appreciation of some of the principles of experimental design.

Interpretation

The final part of the experimental process is interpretation of the results. It is nevertheless considered as part of design, since the criteria on which decisions are made about experimental results should be anticipated. Interpretation is concerned with the conclusions that can be drawn from the results of an experiment. It requires an appreciation of the scientific theory which lies behind the experiment, the response variable, the independent factors and the statistical model. The central part of interpretation is the *decision* about the theory being tested (Lindley 1973). There are no absolute, only relative, criteria for this decision and in some comparative experiments the criteria are based on probabilities provided by statistical models. Whether or not a probabilistic statement of a result is provided, the scientist should have decided in advance what his judgement will be of the possible outcomes of experiments. In non-statistical experiments this might be based on some pattern in data or even on the 'feel' of the scientist. In statistically based experiments judgement is expressed in the value he sets for the significance level at which the null hypothesis is accepted or rejected. That is the point at which the decision about the result of an experiment changes. If there is no decision, it could mean that a critical experiment has not been designed or, more probably, sufficient thought has not been given to the null hypothesis (Strong 1982). However, experimenting is a learning process and generally, whatever the outcome, the scientist is prompted to design further experiments, or to make refinements which provide fuller answers to scientific problems. The approach to interpretation should be within the context of the conjectural character of scientific knowledge.

In order to introduce some of the problems of interpretation it is necessary to use a simple experiment involving use of a statistical model – the t-test. Some basic knowledge of statistics is assumed but descriptions of the test and the rationale underlying it may be found in standard textbooks on elementary statistics. Let us consider a situation where we have measured the density of sediment in 20 samples of water in a river under flood conditions having previously measured the density of sediment in a *very large number* of samples from the same site in the stream in low flow conditions. We want to test the hypothesis that the sediment concentration in flood conditions is greater than in low flow conditions. We test the hypothesis that the mean of the sample of 20 (x) is greater than the mean of the population (X).

Our idea that the concentration of sediment in samples of water from a river in flood conditions is greater than in low flow conditions is based on mechanical theories of flow and sediment entrainment. To test whether this is the case we would use the mean of the 20 sample values (x) and determine the probability that it came from the population of all possible means of the same sample size. The null hypothesis under test is that the sample mean comes from that population of sample means of size 20. Note that we are dealing with a mean value of 20 samples and we can only usefully compare it with other means of sample size 20. We compare it with the sampling distribution of means of sample size 20 and ask the probability that our sample comes from this population of means (Figure 12.4).

The null hypothesis is that the sample mean does come from this population of sample means and we only reject it if the chance of it being true is so low that we judge the alternative hypothesis to be acceptable. Again it is important to emphasize that

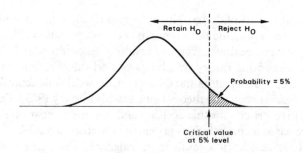

Figure 12.4 One-sided probability: critical value and region of rejection of H_0.

the decision is not absolute. The level chosen for making this decision is a matter of practice and judgement. Since we mention direction in our hypothesis (greater than) we use a one-tail test. Let us say that the one-sided probability of us achieving a value of the mean of 20 samples as large as that obtained, assuming it came from the same population, is found to be 0.065. This means that if flood conditions had no effect on sediment concentration, there would be a 0.065 chance of obtaining a mean of 20 samples as large or larger than the one we actually measured.

Classical statistical methods involve the extra notion of acceptance or rejection of the null hypothesis, by introducing a level of significance, which can be called a credibility index for the null hypothesis. Assuming the null hypothesis to be true, we state what we can expect of an individual mean drawn from a population. We set the limit to our tolerance of the difference between sample and population means before we consider the possibility that the sample comes from a different population. Usually this is expressed as some simple, small proportion of the population of means which lies at such a distance from the overall mean that we decide that if a value is actually found in this region, we will consider it not in fact to have come from the population, but from some other. This region of rejection is the significance level. Usually, with this region of rejection goes a critical value of a statistic against which we compare our measured value.

In this decision-making process we run the risk of making two types of error. Consider Figure 12.5. The first part shows what the world would be like if the null hypothesis were true. We choose a level of significance α of 5%. There is a finite (5%) chance of observing a mean in the shaded area and thus rejecting the null hypothesis.

Rejecting the null hypothesis when it is in fact true is called a type I error. This amounts to accepting our working hypothesis that the sample mean is greater than the population mean when it is in fact false. We would have accepted a theory as being true when it was not. This is regarded as the most serious error and the workings of statistics minimize this error by setting α sufficiently small.

Suppose, however, that the null hypothesis is false as in the second part of Figure

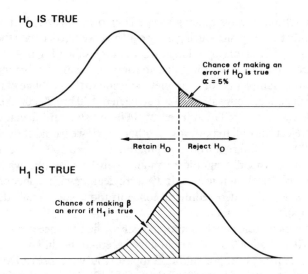

Figure 12.5 Type I and Type II errors

12.5. All the possible sample means have a different distribution. In this situation the correct decision would be to reject the null hypothesis. Of all the possible values of the means from the second distribution a number would fall in the shaded region where we would accept the null hypothesis. Again we would have made a mistake, that is, accepting the null hypothesis when it was false. This type of error is called a type II error. It amounts to not accepting a theory when it is in fact true. This is regarded as a less serious error.

Statistical practice distinguishes between type I and type II errors in that statistical tests are designed to minimize the former. This objective is consistent with the critical approach to science. By saying that a theory is accepted as true, the statistician really means that it stands for the time being as a better theory than the null hypothesis. Theories supported by experiments remain to be tested by new experiments. Accepting a theory as true when it is false (a worse theory than the null hypothesis), the type I error, amounts to deciding that no further tests are necessary. This is the grave mistake that statistics tries to avoid. To accept a theory as false when it is true (a better theory than the null hypothesis), the type II error, is a less serious error because the theory still lives to be reformulated and a new test situation can be devised. Thus, the scientist's response to the test situation is to have a tendency to reject theories, since there is always another chance for the theory to show its mettle.

It is easy to see that if we decrease the chance of making a type I error by decreasing α then we increase the chance of making a type II error. Using a legal analogy, by making our legal procedures less likely to convict an innocent man we make it more likely that guilty men will go free. Further we can never eliminate the chance of making a type I error without always making a type II error. However, just as we can improve our legal procedure by introducing more evidence we can decrease the chances of both type I and type II errors by taking larger samples (see Figure 12.6) which reduce the standard error of the sampling distributions of means. Unfortunately, the use of larger and larger samples is not a complete solution to the problem of type I and type II errors. For instance, if our sample size had been not 20 but 20,000 then a value of the mean of 1002 with a population mean of 1000 and standard deviation of 100 would allow us to reject the null hypothesis. In other words we would conclude that a sample of mean 1002 was different from a population of mean 1000. Whether or not such a difference is statistically significant might then become irrelevant since it might not have any physical significance. If we were dealing with a phenomenon as inherently variable as suspended sediment concentration, such small differences would have no importance whatsoever.

We are led to the inevitable conclusion that levels of significance are arbitrary. If we had set α at 0.10 instead of 0.05 we would have accepted the difference as significant. And the problem is yet more subtle. Suppose you were presented with the results of 100 statistical tests, and you found that, of the 100, 10 were significant, with α as 0.05. How would you interpret them? Would you consider that in 10 out of the 100 situations there were real effects that produced significant test statistics? Or would you consider the probability of getting 10 significant test statistics out of 100 at the 0.05 level? The chance of getting so many significant statistics at the 0.05 level is near 50%. In other words, with a significance level of 0.05 there is a 50% chance of getting 10 significant test statistics out of 100 even when there is no real effect, and yet

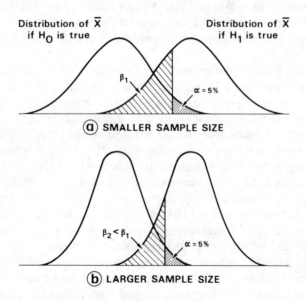

Figure 12.6 Effect of sample size on type I and type II errors

each significant result might be treated as a real effect. At a 0.05 level with 100 cases we would *expect* 5 significant results. The situation where we have many test statistics is very common and the problem of collections of test statistics needs to be considered over and above individual values.

The problem of significance is still deeper than shown by these straightforward examples. As in our example, what if the mean lay just outside the region of rejection? The probability of getting a mean as large as the one we got was 0.065 whereas the significance level was 0.05 and so we decided to retain the null hypothesis of no difference. However, our working hypothesis is of a greater sediment concentration in flood conditions. What this means is that the empirical evidence supports our theory but is in fact used to reject it. This might seem particularly unfortunate if we have prior grounds, that is independent evidence, for believing our theory. On this basis, experimental and statistical methods seem designed to protect the null hypothesis (Lindley 1973; Strong 1982). They seem designed to prevent us from accepting any theories. And some may argue that the null hypotheses are merely responses to our working hypotheses, they are merely 'straw men' designed to be knocked down, and results such as the one in our example should count as supporting the working hypothesis even though they are 'not significant'.

The basic defence of the statistical method is to reemphasize that the level of significance is chosen by the experimenter. If the method seems too protective of the null hypothesis, then the level of significance can be changed. If the level of α appears arbitrary, then it must be recognized that all levels are arbitrary but some level must be chosen. If the results are unsatisfactory, then this must mean that there is some independent evidence to show that this is so. And if this is the case, a new experiment can be designed which takes into account this evidence.

The real cause of such problems lies in poor appreciation of objects or systems and poor design of experiments in the first place. If the interpretation, based on some particular level of significance, seems inadequate, then the question arises as to why that particular level was chosen in the first place. This is not to argue against criticism of experiments and for rejection of significance levels, which are inevitable, but against the ad hoc manipulation of significance levels to produce particular results.

Conclusions

The response to uncertainty in deciding about the results of experiments should be twofold. First, the attitude of mind of the enquiring, critical scientist, which does not seek to support theories, should be adopted. 'The central mistake . . . is *the quest for certainty*' (Popper 1972b, p. 63). Secondly, there should be a constant effort to improve our experimental design by careful choice of variables, statistical models and sampling programme by treating the null hypothesis as a competing hypothesis and exploring the consequences of its validity as the bases for designing a properly critical experiment between competing hypotheses. Further, the possible outcomes of our experiment should be considered with their interpretation. If the results of an experiment are unsatisfactory or inconsequential then the scientist should *improve the experiment* or *design a new one*. All experiments are open to criticism and improvement in the same way that theories are.

A critical test of your knowledge of experimental design can be undertaken by attempting Exercise 12.2.

CHAPTER 13:
PHYSICAL GEOGRAPHY AND THE
CRITICAL TRADITION

'. . . you must murder your darlings'

Two assumptions lie behind the arguments and ideas of the preceding chapters. The first is that, in physical geography as a whole, there are serious problems in the way it is conducted. Such misgivings are shared by others (Chorley 1971; Clayton 1973; Kennedy 1977, 1980; Dury 1978) if not all for the same reasons. The problems referred to cannot be stated in brief and simple terms because in analysing the structure of the subject there are relations between problems, theories, methods and interpretation which are difficult to disentangle. The main part of the book pursues this analysis particularly through the use of case studies. Nevertheless, what is clear is that in physical geography there have been very few advances in our theories about, or our understanding of, the natural world. In parallel disciplines, such as geology, ecology and meteorology, the past few decades have witnessed major changes in theories, many discoveries about the structure and behaviour of natural systems, and fundamental conceptual changes. Some of these important advances are illustrated in this book. In one sense these advances belong to physical geography, since we share in them, but it would be foolish to claim that they are advances of physical geography. The discipline, as it is taught and identified by professional physical geographers, can boast no major advances. In addition, the vast majority of journals and advanced texts still contain material which is either merely descriptive or an attempt to model some phenomenon by statistical or simple mathematical equations akin to those employed by engineers.

The second assumption which lies behind the arguments of this book is that any advance in the subject can only follow from a change in its traditions which embodies improvements in method. It would be absurd to suggest that in order to be effective a scientist *must* have a particular view of science, or be able to identify the logical structure of his argument, or follow a particular sequence of actions in experimentation. It is also absurd to suggest that he must be aware of the philosophy of science in order to do good science. However, the advances which have been made in science have involved theorizing, experimentation, argument and above all criticism. They have done so because these things are in the *scientific tradition*. It is this tradition which physical geography lacks. If it is to advance, increasing our understanding of the world, then a tradition must be established by a conscious effort to inject an awareness of method and to improve the methods we use. This tradition is expressed by Arthur Holmes in the preface to *Principles of Physical Geology* (1965):

Every chapter contains exciting stories of man's achievements and speculations, tempered by the sobering reflection that while we may often be wrong, Nature cannot be. It may be noticed that inconsistencies have arisen here and there, when certain problems are approached along different lines. Such difficulties are not glossed over. By bringing them into the open, instead of sweeping them under the carpet of outworn doctrines and traditional assumptions, the reader is helped to see where further research is needed and may even be stimulated to take part in it. (p. vii)

The argument throughout the preceding chapters is that an awareness of method, the shaping of ideas and the logic of argument all have great *practical* implications. Studying method is not an esoteric, marginal activity of science. Although it is not possible to state that it is absolutely necessary, it is, we believe, something every scientist should do. There may be the occasional chance discovery or the rare genius for whom such advice may seem irrelevant, but even chance discoveries and the work of geniuses are based on the firm foundations of the achievements of those who have gone before. These firm foundations are to be achieved through a tradition of good scientific method. This method involves skills of thinking and doing which can be acquired and improved. There are some important parts of the scientific process, such as introducing new ideas and concepts, which are not rational and cannot be taught. But critically testing theories and recognizing problem situations are rational processes and can be developed as skills.

A corollary of the arguments presented is the view we take of the history of development of physical geography. It is a misconception that the subject has become more scientific in recent years. For geomorphology in particular, there is an almost universal misunderstanding that, with the retreat from the Davisian paradigm and the subsequent emphasis on processes and quantification, it has become a science. In fact a perusal of the leading British journal of geomorphology will show that, if anything, geomorphology is developing into a minor branch of engineering. The significance of this is that the engineer is more preoccupied with successful modelling and prediction than with explanation or truth. The changes which have occurred have been a retreat from the historically important problems about landscapes and landforms to problems of modelling but without the necessity to do so provided by the testing of theories. In some branches of the subject, such as Quaternary studies, there remains an awareness of a set of problems and theories which are characteristic and have an ancestry within the subject. In physical geography in general the trend away from geomorphological, biogeographical or other theory to the solution of engineering problems or to the application of the theories of other disciplines is, however, clear (cf. SRE 1977).

The major changes which took place from the 1950s onwards can be regarded as shifts of interest and developments of technique. These changes are of little consequence when considering whether or not the subject has developed as a science and whether or not our understanding of the world has grown. In terms of theory and understanding, quantitative measures, statistical models and sophisticated apparatus have in themselves little to do with science. Nevertheless, the changes which have taken place have had an impact on physical geography since they affect the questions physical geographers ask. There is nothing in this situation to criticize since the source of a question is of no scientific importance. We cannot condemn the lucky person who wants to study a problem of regional hydrology because he has just acquired a system for digital image processing on the grounds that he is not being scientific. But this situation is disappointing because there is no theoretical progression from any problem and no speculation is provided by equipment. And this is what has happened in physical geography. The theoretical framework seems not to have developed as in other disciplines. Consider the amused fascination with which we read the accounts of Leonardo da Vinci, Du Buat, Playfair or Gilbert on rivers and fluvial landscapes (Chorley et al. 1964). It is uncanny how modern they appear to be. Do they reveal what little advance has been made? It would appear that the so-called revolutions in physical geography are not revolutions in the Kuhnian sense, however liberally we interpret those ideas. The changes which have taken place can be interpreted rather as changes in fashion. Thus there has been change but little real progress.

If physical geography is to progress, then the question arises as to what should be studied. Medawar's answer to this problem in his *Advice to a Young Scientist* (1979) is to find out what people are arguing about. Perhaps this does not go far enough. The point about physical geography is that the bases for proper scientific argument are not well established. There need to be developments in theorizing, experimentation and the ability to recognize problems for progress to be achieved. In particular we must develop the tradition of criticism. As Popper has put it:

> ... my answer to the question 'How do you know? What is the source or the basis of your assertion? What observations have led you to it?' would be: 'I do not know: my assertion was merely a guess. Never mind the source, or the sources, from which it may spring – there are many possible sources and I may not be aware of half of them; and origins and pedigrees have in any case little bearing on truth. But if you are interested in the problem which I tried to solve by my tentative assertion, you may help me by criticising it as severely as you can; and if you can design some experimental test which you think might refute my assertion, I shall gladly, and to the best of my power, help you to refute it. (1972a, p. 27)

Or, as Noel Coward wrote:

> ... you must murder your darlings.

EXERCISES

The exercises are numbered in relation to the relevant chapter which provides a context and background information.

1.1 Consider the explanation given for the origin of arroyos in Case Study 1.3. Try to suggest as many explanations for the features as you can using different covering laws and statements of initial conditions. How would you decide between these competing explanations?

2.1 Consider the area shown in Plate I. Using the approach envisaged by the classical tradition, describe what observations you might make in the field that would eventually allow you to give an account of the physical geography of the area.

3.1 Study the area shown in Plate I again. How would you test the proposition that the features shown in the picture are alluvial terraces? What competing hypotheses might be advanced to explain the bench-like features? What field data would you collect that would enable you to test each theory critically? (see Sissons and Cornish 1983)

4.1 Consider the theory of plate tectonics described in Case Study 4.1. Follow up any other accounts you can find of the theory and the way it gained acceptance over the view that the earth's surface was stable. Try to present alternative accounts of the case of plate tectonics using the outlook of someone who maintains:

(i) the classical view of science;

(ii) the critical rationalist view of science.

Which of the accounts do you find most acceptable?

4.2 Choose any subject area of the environmental sciences that is familiar to you. To what extent do you consider that its history corresponds to Kuhn's model?

4.3 A major contention of Kuhn's model is that comparison of theories across paradigms or different disciplinary matrices is difficult if not impossible because of the problem of incommensurability. Consider this thesis and:

(i) examine the 'scientific revolutions' described in Case Studies 4.1 and 4.2, and see if you can discover any features of the pairs of competing paradigms which suggest this idea of incommensurability;

(ii) looking at the history of the matter, say to what extent you would agree with the assertion that the switch represented by each revolution was not based upon evidence but rather, something more like psychological or sociological factors.

5.1 To what extent would you go along with the suggestion that the ideas described in Case Studies 5.1 and 5.2 represent distinctive research programmes in the sense described by Lakatos? Do you consider these ideas are better described as paradigms in the sense described by Kuhn? Can you think of any better way of describing the scientific work in these case studies than in terms of the ideas of Lakatos or Kuhn?

6.1 Consider the experience of Ruse (1982) at the monkey trial. In the context of evolutionary theory do you consider that the dictum 'anything goes' is a good one?

7.1 Popper has written extensively on the scientific character of Darwin's theory of evolution (see Popper 1972b, pp. 267–270; 1974, pp. 105–109; and especially 1976, pp. 167–180). His views have been interpreted in terms of suggesting that Darwin's theory is not scientific. The character of the debate which has subsequently taken place can be seen in the papers of Peters (1976), Brady (1979, 1982), Halstead (1980), Little (1980), Ruse (1981), Ridley (1981), Sparks (1981), and Charlesworth (1982). Although Popper (1980) and others (e.g. Bondi 1980) have denied that the scientific character of evolutionary theory is in question, the arguments are relevant to those interested in historical sciences like geology, geomorphology and evolutionary biology.

Review the scientific status of the theory of evolution. What implications do you draw for the various accounts of the methods of the natural sciences?

8.1 For any published study in physical geography which is familiar to you try to identify a theory which is being tested. Try to specify any meta-theory or myth which underlies the theory being tested. Consider the implications of a failure of the theory for these other ideas. On what basis could these other ideas be rejected?

9.1 Consider the models described in Case Studies 9.1 and 9.2 and May's (1975, 1981) models of ecosystem development. In the testing of the predictions from each of these models specify the possible sources of discrepancy between prediction and observation. How would you judge the actual source of discrepancy?

10.1 Look again at Case Studies 8.1 and 8.2, and consider them in the light of Murdoch's (1966) analysis.

Is the material of Case Studies 9.1 and 9.2 susceptible to the same criticism?

11.1 Define variables on the four scales of measurement which represent properties of each of the following: rivers, morphogenetic regions, soils, climate, moraines. For pairs of variables which you can relate theoretically specify the appropriate statistical relation which will provide a test of the theory.

11.2 For a problem in physical geography which interests you find a recent paper which reports the results of research. By following up the material to which it makes reference try to establish the ancestry of the ideas it contains.

12.1 (a) Search the literature for examples of measured relation and assess the extent to which randomization has been and should have been employed in experimental design.

(b) You are given the problem of examining the relation between altitude and erosion rates on granite with the notion that they are inversely related.

You are given three response variables:
−lowering of exposed rock surfaces over time
−depth of weathered mantle
−solute losses from drainage basins.

For each response variable speculate on:
(i)   the extraneous factors which affect them;
(ii) how you would control each of these factors;
(iii) the method of randomization you would employ.

12.2 Speculate on each of the hypothetical experimental problems given below. For each give details of variables, controls of extraneous variables, layout of the experiment, sampling design, method of randomization and possible interpretation of results.

(a) Using a climatic simulator with full control over temperature and humidity, design an experiment to test the relation between freeze–thaw cycle frequency and rock disintegration.

(b) Design an experiment to find the effect of valley topography on airflow patterns.

(c) Design an experiment to assess the impact of the Mount St Helens eruptions.

Plate 1 Panorama showing upper Glen Roy, Scotland. Photograph looks WSW from a site above the Falls of Glen Roy (OS Map Ref. NN 360927).

REFERENCES

Abler, R., Adams, J.S. and Gould, P. (1971) *Spatial Organisation: the Geographers View of the World* (Prentice Hall: Hemel Hempstead)

Andersen, J.L. and Sollid (1971) Glacial chronology and glacial geomorphology of the glaciers Midtdalsbreen and Nigardsbreen, southern Norway, *Norsk Geografisk Tidsskrift* 25: 1–38

Anderson, M.G. and Burt, T.P. (1981) Methods of geomorphological investigation, in *Geomorphological Techniques*, A. Goudie (George Allen and Unwin: London) pp. 1–11

Ayer, A.J. (1950) Basic propositions, in *Philosophical Analysis*, M. Black (Cornell University Press: Ithica) pp. 60–74

Ayer, A.J. (1964) *Language Truth and Logic*, second edition (Victor Gollancz: London)

Bagnold, R.A. (1966) An experimental approach to the sediment transport problem from general physics, *United States Geological Survey, Professional Paper*: 282 E 422–421

Baker, N.V. (1978) The Spokane Flood controversy, in *The Channeled Scablands*, N.V. Baker and D. Nummedal (NASA: Washington DC) pp. 3–6

Baker, N.V. and Payne, S. (1978) G.K. Gilbert and modern geomorphology, *American Journal of Science* 278: 97–123

© Baker, V.R. and Bunker, R.C. (1985) Cataclysmic late Pleistocene flooding from Lake Missoula: a review, *Quarternary Science Review*, 4: 1–43

Barnes, B. (1982) *T.S. Kuhn and Social Science* (Macmillan: London)

Battarbee, R.W., Flower, R.J., Stevenson, J. and Rippey, B. (1985) Lake acidification in Galloway: a palaeological test of competing hypotheses, *Nature* 314: 350–352

Bird, J.H. (1975) Methodological implication for geography from the philosophy of K.R. Popper, *Scottish Geographical Magazine* 9: 153–163

Bishop, P. (1980) Popper's principle of falsifiability and the irrefutability of the Davisian cycle, *Professional Geographer* 32: 310–315

Biswas, A.K. (1970) *History of Hydrology* (North Holland Publishing Co: Amsterdam)

Bondi, H. (1980) Evolution, *New Scientist* 87: 671

Bonney, T.G. (1895) *Charles Lyell and Modern Geomorphology* (Macmillan: London)

Bradley, W.H. (1962) Geologic laws, in *The Fabric of Geology*, C.C. Arbritton (Addison-Wesley: Reading, Mass.) pp. 12–23

Brady, R.H. (1979) Natural selection and the criteria by which a theory is judged, *Systematic Zoology* 28: 600–621

Brady, R.H. (1982) Dogma and doubt, *Biological Journal of the Linnean Society* 17: 78–96

Bretz, J.H. (1923) Channeled Scablands of the Columbia Plateau, *Journal of Geology* 31: 61–649

Bretz, J.H. (1978) Introduction, in *The Channeled Scablands*, N.V. Baker and D. Nummedal (NASA: Washington DC) p. 1.

Brunsden, D. and Thornes, J.B. (1979) Landscape sensitivity and change, *Institute of British Geographers, Transactions* NS4: 463–484

Brush, L.M. and Wolman, M.G. (1960) Knick-point behaviour in non-cohesive materials: a laboratory study, *Geological Society of America Bulletin* 71: 59–74

Burke, T.E. (1983) *The Philosophy of Popper* (Manchester University Press: Manchester)

Carson, M.A. (1981) Influence of porefluid salinity on instability of marine clays: a new approach to an old problem, *Earth Surface Processes and Landforms* 6: 499–515

Chalmers, A.F. (1982) *What Is This Thing Called Science?* second edition (Open University Press: Milton Keynes)

Chamberlin, T.C. (1890) The method of multiple working hypotheses, *Science* XV: 92–96

Chapman, K. (1981) Issues in environmental impact assessment, *Progress in Human Geography* 5: 198–210

Charlesworth, B. (1982) Neo-Darwinism – the plain truth, *New Scientist* 94: 133–137

Chorley, R.J. (1962) Geomorphology and General Systems Theory, *United States Geological Survey, Professional Paper* 500B

Chorley, R.J. (1966) The application of statistical methods to geomorphology, in *Essays in Geomorphology*, G.H. Dury (Methuen: London) pp. 77–100

Chorley, R.J. (1967) Models in geomorphology, in *Models in Geography*, R.J. Chorley and P. Haggett (Methuen: London) pp. 59–90

Chorley, R.J. (1971) The role and relations of physical geography, *Progress in Physical Geography* 2: 87–109

Chorley, R.J. (1978) Bases for theory in geomorphology, in *Geomorphology: Present Problems and Future Prospects*, C. Embleton, D. Brunsden and D.K.C. Jones (Oxford University Press: Oxford) pp. 1–13

Chorley, R.J., Dunn, A.J. and Beckinsale, R.P. (1964) *The History of the Study of Landforms or the Development of Geomorphology*, Volume 1 (Methuen and John Wiley: London)

Clayton, K.M. (1970) The problems of field evidence in geomorphology, in *Geographical Essays in Honour of K.C. Edwards*, R.H. Osborne, F.A. Barnes and J.C. Doornkamp (Department of Geography, University of Nottingham: Nottingham) pp. 131–139

Clayton, K.M. (1973) Second International Geographical Congress, Montreal, August 1972, *Zeitschrift für Geomorphologie* NF17: 246–249

Cole, L.C. (1960) Competitive exclusion, *Science* 132: 348–349

Colinvaux, P.A. (1973) *Introduction to Ecology* (John Wiley and Sons: New York)

Cooke, R.U. and Reeves, R.W. (1976) *Arroyos and Environmental Change in the American South-West* (Clarendon Press: Oxford)

Cox, D.R. (1958) *Planning of Experiments* (Wiley: New York)

Darwin, C. (1872, reprinted 1958) *Origin of Species*, sixth edition (Mentor: London)

Darwin, F. (1903) *More Letters of Charles Darwin* (Murray: London)

Davis, W.M. (1921) Features of glacial origin in Montna and Idaho, *Annals of the Association of American Geographers* 10: 122–132

Davis, W.M. (1926) The value of outrageous geological hypotheses, *Science* 63: 463–468

Derbyshire, E., Gregory, K.J. and Hails, J.R. (1979) *Geomorphological Processes* (Dawson: Folkestone)

Diamond, J.M. and May, R.M. (1981) Island biogeography and the design of nature reserves, in *Theoretical Ecology*, R.M. May, second edition (Blackwell Scientific Publications: Oxford) pp. 228–253

Du Buat, L.G. (1786) *Principles d'Hydraulique*, Volume 1 (Paris)

Dunne, T. and Black, R.D. (1970a) An experimental investigation of runoff production in permeable soils, *Water Resources Research* 6: 478–490

Dunne, T. and Black, R.D. (1970b) Partial-area contribution to storm runoff in a small New England watershed, *Water Resources Research* 6: 1286–1311

Dunne, T. and Leopold, L.B. (1979) *Water in Environmental Planning* (W.H. Freeman: San Francisco)

Dury, G.H. (1978) The future of geomorphology, in *Geomorphology: Present Problems and Future Prospects*, C. Embleton, D. Brunsden and D.K.C. Jones (Oxford University Press: Oxford) pp. 263–274

Dury, G.H. (1980) Neo-catastrophism, a new look, *Progress in Physical Geography* 4: 391–413

Edwards, K.J. (1983) Quaternary palynology: a consideration of a discipline, *Progress in Physical Geography* 7: 113–125

Eliot, T.S. (1975) *Collected Poems, 1909–1962* (Faber: London)

Feyerabend, P. (1970) Consolations for the specialist, in *Criticism and the Growth of Knowledge*, I. Lakatos and A. Musgrave (Cambridge University Press: Cambridge) pp. 197–230

Feyerabend, P. (1975) *Against Method* (Verso: London)

Feyerabend, P. (1978) *Science in a Free Society* (Verso: London)

Feyerabend, P. (1981) How to defend society against science, in *Scientific Revolutions*, I. Hacking (Oxford University Press: Oxford) pp. 156–167

Fisher, R.A. (1966) *The Design of Experiments*, eighth edition (Oliver and Boyd: London)

Fredrikson, R.L. (1972) Nutrient budget of a Douglas-Fir forest on an experimental watershed in western Oregon, in *Proceedings-Research on Coniferous Forest Ecosystems – a Symposium* (Bellingham: Washington) pp. 115–131

Freeze, R.A. (1972) Role of subsurface flow in generating surface runoff. 2–upstream source areas, *Water Resources Research* 8: 1272–1282

Gerrard, A.J. (1978) Tors and granite landforms of Dartmoor and Bodmin Moor, *Proceedings of the Usher Society* 42: 94–100

Gerrard, A.J. (1984) Multiple working hypotheses and equifinality: a reply, *Transactions of the Institute of British Geographers* NS9: 367–371

Gilbert, F.S. (1980) The equilibrium theory of island biogeography: fact or fiction? *Journal of Biogeography* 7: 209–235

Gilbert, G.K. (1896) The origin of hypotheses, illustrated by the discussion of a topographical problem, *Science* 3: 1–13

Gillespie, C.C. (1959) *Genesis and Geology* (Harper and Row: New York)

Goodman, D. (1975) The theory of diversity–stability relationships in ecology, *Quarterly Review of Biology* 50: 237–266

Gould, P. (1981) Letting the data speak for themselves, *Annals of the Association of American Geographers* 71: 166–176

Graf, W. (1979) The development of montane arroyos and gullies. *Earth Surface Processes* 4: 1–14

210 References

Grass, A.J. (1970) Initial instability of fine bed sand, *Journal of the Hydraulics Division American Society of Civil Engineers* 96: 619–632

Grass, A.J. (1971) Structural features of turbulent flow over smooth and rough boundaries, *Journal of Fluid Mechanics* 50: 233–255

Gregory, D. (1978) *Ideology, Science and Human Geography* (Hutchinson: London)

Griffey, N.J. and Matthews, J.M. (1978) Major Neoglacial expansions in southern Norway: evidence from moraine ridge stratigraphy with ^{14}C dates in buried paleosols and moss layers, *Geografiska Annaler* 60A: 73–90

Grove, J.M. (1972) The incidence of landslides, avalanches and floods in western Norway during the Little Ice Age, *Arctic and Alpine Research* 4: 131–138

Gruber, H.E. (1974) *Darwin on Man* (Wildwood House: London)

Guelke, L. (1971) Problems of scientific explanation in geography, *Canadian Geographer* XV: 38–53

Hack, J.T. (1960) Interpretation of erosional topography in humid temperate regions, *American Journal of Science, Bradley Volume* 258–A: 80–97

Hacking, I. (1981) Lakatos's philosophy of science, in *Scientific Revolutions*, I. Hacking (Oxford University Press: Oxford) pp. 128–143

Hacking, I. (1983) *Representing and Intervening* (Cambridge University Press: Cambridge)

Haggett, P. and Chorley, R.J. (1967) Models, paradigms and the new geography, in *Models in Geography*, R.J. Chorley and P. Haggett (Methuen: London) pp. 19–42

Hairston, N.G., Smith, F.E. and Slobodkin, L.B. (1960) Community structure, population control and competition, *American Naturalist* 94: 421–425

Hallam, A. (1972) Continental drift and the fossil record, *Scientific American* 227: 56–66

Hallam, A. (1973) *A Revolution in the Earth Sciences* (Clarendon Press: Oxford)

Halstead, B. (1980) Popper: good philosophy or bad science? *New Scientist* 87: 215–217

Hamilton, T.H., Rubinoff, I., Barth, R.H. and Bush, G.L. (1963) Species abundance: natural regulation of insular variation, *Science* 142: 1575–1577

Harary, F. (1971) What is and what is not a mathematical model, *Biometric Praximetrie* XII 1–4: 1–16

Harbaugh, J.W. and Bonham-Carter, G. (1970) *Computer Simulations in Geology* (Wiley: New York)

Hardin, G. (1960) The competitive exclusion principle, *Science* 131: 1292–1297

Harré, F. (1972) *The Philosophies of Science* (Oxford University Press: Oxford)

Harvey, D. (1969) *Explanation in Geography* (Edward Arnold: London)

Hawksworth, D. and Rose, F. (1976) *Lichens as Air Pollution Monitors* (Edward Arnold: London)

Hemple, C.G. (1966) *Philosophies of Natural Science* (Prentice-Hall: New Jersey)

Hewlett, J.D. and Hibbit, A.R. (1963) Moisture and energy conditions within a sloping soil mass during drainage, *Journal of Geophysical Research* 68: 1081–1087

Hewlett, J.D. and Hibbert, A.R. (1967) Factors affecting the response of small watersheds to precipitation in humid regions, in *International Symposium on Forest Hydrology*, W.E. Soppar and H.W. Lull (Pergamon Press: London) pp. 275–290.

Hicks, C.R. (1973) *Fundamental Concepts in the Design of Experiments* (Holt Reinhart Winston: New York)

Holmes, A. (1965) *Principles of Physical Geography*, second edition (Nelson: London)

Holt-Jensen, A. (1981) *Geography: Its History and Concepts: A Student Guide* (Harper and Row: London)

Holton, G. and Roller, D.H.D. (1958) *Foundations of Modern Physical Sciences* (Addison-Wesley: Reading, Mass.)

Horn, H.S. (1981) Succession, in *Theoretical Ecology*, R.M. May, second edition (Blackwell Scientific Publications: Oxford) pp. 253–271

Hospers, J. (1980a) What is explanation? in *Introductory Readings in the Philosophy of Science*, E.D. Klemke, R. Hollinger and A.D. Kline (Prometheus Books: New York) pp. 87–103

Hospers, J. (1980b) Law, in *Introductory Readings in the Philosophy of Science*, E.D. Klemke, R. Hollinger and A.D. Kline (Prometheus Books: New York) pp. 104–111

Hutchinson, J.N. and Bhandari, R.K. (1971) Undrained loading: a fundamental mechanism of mudflows and their mass movement, *Geotechnic* 21: 353–358

Hutton, J. (1795) *Theory of the Earth*, 2 volumes (Edinburgh)

Jeffers, J.N.R. (1978) *Statistical Checklist 1* (Institute of Terrestrial Ecology, Natural Environmental Research Council: London)

Johnson, M.P. and Raven, P.H. (1973) Species number and endemism, *Science* 179: 893–895

Johnston, R.J. (1978) Paradigms and revolutions or evolution: observations on human geography since the Second World War, *Progress in Human Geography* 2: 189–206

Johnston, R.J. (1979) *Geography and Geographers* (Arnold: London)

Johnston, R.J. (1983) *Philosophy and Human Geography* (Arnold: London)

Jones, B. (1974) Plate tectonics: a Kuhnian case? *New Scientist* 63: 536–538

Joyce, J. (1960) *Ulysses* (Bodley Head/Penguin Edition: London)

Keat, R. and Urry, J. (1975) *Social Theory as Science* (Routledge and Kegan Paul: London)

Kennedy, B.A. (1977) A question of scale? *Progress in Physical Geography* 1: 154–157

Kennedy, B.A. (1980) A naughty world, *Institute of British Geographers Transactions* NS4: 550–558

Kennedy, B.A. (1983) An outrageous hypotheses in geography, *Geography* 68: 326–330

Kennedy, B.A. (1984) On Playfair's law of accordant junctions, *Earth Surface Processes and Landforms* 9: 153–173

Kitts, D.B. (1962) The theory of geology, in *The Fabric of Geology*, C.C. Albritton (Addison-Wesley: Reading, Mass) pp. 49–68

Klein, M. (1976) Hydrograph peakedness and basin area, *Earth Surface Processes* 1: 27–30

Krumbein, W.C. (1959) The sorting of geological variables illustrated by regression analysis of factors controlling beach firmness, *Journal of Sedimentry Petrology* 29: 575–587

Krumbein, W.C. (1960) Stratigraphic maps from data observed at outcrop, *Proceedings of the Yorkshire Geological Society* 32: 353–366

Kuhn, T.S. (1962) *The Structure of Scientific Revolutions*, first edition (University of Chicago Press: Chicago)

Kuhn, T.S. (1970a) *The Structure of Scientific Revolutions*, second edition (University of Chicago Press: Chicago)

Kuhn, T.S. (1970b) Logic of discovery or the psychology of research? in *Criticism and the*

Growth of Knowledge, I. Lakatos and A. Musgrave (Cambridge University Press: Cambridge) pp. 1–24

Kuhn, T.S. (1977) *The Essential Tension* (University of Chicago Press: Chicago)

Lacey, A.R. (1976) *A Dictionary of Philosophy* (Routledge and Kegan Paul: London)

Lacy, O. (1953) *Statistical Methods in Experimentation* (Macmillan: London)

Lakatos, I. (1970) Falsification and the methodology of scientific research programmes, in *Criticism and the Growth of Knowledge*, I. Lakatos and A. Musgrave (Cambridge University Press: Cambridge) pp. 91–196

Lakatos, I. (1981) History of science and its rational reconstruction, in *Scientific Revolutions*, I. Hacking (Oxford University Press: Oxford) pp. 107–127

Leeder, M. (1982) *Sedimentology* (Allen and Unwin: London)

Leopold, L.B. and Maddocks, T. (1953) The hydraulic geometry of stream channels and some physiographic implications, *US Geological Survey Professional Paper* 252: 1–56

Leopold, L.B., Wolman, W.G. and Miller, J.P. (1964) *Fluvial Processes in Geomorphology* (W.H. Freeman: San Francisco)

Lindley, D. (1973) *Making Decisions* (Wiley Interscience: London)

Linton, D.L. (1955) The problem of tors, *Geographical Journal* 121: 470–487

Little, J. (1980) Evolution: myth, metaphysics or science? *New Scientist* 87: 708–709

Locke, W.W., Andrews, J.T. and Webber, P.J. (1979) *A Manual of Lichenometry*, British Geomorphological Research Group Technical Bulletin, 26

Lyell, C. (1833) *The Principles of Geology*, Volume III (John Murray: London)

Makin, J.H. (1962) Rational and empirical methods of investigation in geology, in *The Fabric of Geology*, C.C. Albritton (Addison-Wesley: Reading, Mass.) pp. 134–163

Mandeville, A.N., O'Connell, P.E., Sutcliffe, J.V. and Nash, J.E. (1970) River flow forecasting through conceptual models. Part III – the Ray catchment at Grendon Underwood, *Journal of Hydrology* 11: 109–128

Marshall, J.U. (1982) Geography and critical rationalism, in *Rethinking Geographical Enquiry*, D.J. Wood (Geographical Monographs, York University, Atkinson College: Canada) pp. 75–171

Masterman, M. (1970) The nature of a paradigm, in *Criticism and the Growth of Knowledge*, I. Lakatos and A. Musgrave (Cambridge University Press: Cambridge) pp. 59–90

Mather, K. (1964) *Statistical Analysis in Biology*, fifth edition (Methuen: London)

Matthews, J.A. (1974) Families of lichenometric dating curves from the Storbreen gletscher-vorfeld, Jotunheimen, Norway, *Norsk Geografisk Tidsskrift* 28: 215–235

Matthews, J.A. (1975) Experiments on the reproducibility of lichenometric dates, Storbreen gletschervorfeld, Jotunheimen, Norway, *Norsk Geografisk Tidsskrift* 29: 97–109

Matthews, J.A. (1977) A lichenometric test of the 1750 end-moraine hypothesis: Storbreeen gletschervorfeld, southern Norway, *Norsk Geografisk Tidsskrift* 31: 129–136

Matthews, J.A. (1980) Some problems and implications of ^{14}C dates from a podsol buried beneath an end moraine at Haugabreen, southern Norway, *Geografiska Annaler* 62A: 185–208

Matthews, J.A. and Dresser, P.Q. (1983) Intensive ^{14}C dating of a buried palaeosol horizon, *Geologogiska Foreningens i Stockholm Forhandlinger* 105: 59–63

May, R.M. (1975) *Stability and Complexity in Model Ecosystems*, second edition (Princeton University Press: Princeton)

May, R.M. (1981) Patterns in a multi-species community, in *Theoretical Ecology*, R.M. May, second edition (Blackwell Scientific Publications: London)

Medawar, P.B. (1963) Is the scientific paper a fraud? *The Listener* LXX: 377–378

Medawar, P.B. (1979) *Advice to a Young Scientist* (Harper and Row: New York)

Melloy, G. (1873) *Geology and Revelation: or the Ancient History of the Earth Considered in the Light of Geological Facts and Revealed Religion* (Burns, Oats and Co: London)

Miller, M.C., McCave, I.N. and Komar, P.D. (1977) Threshold of sediment motion under unidirectional currents, *Sedimentology* 24: 507–528

Miroff, B. (1976) *Pragmatic Illusions: Presidential Politics of J.F. Kennedy* (David McKay: New York)

Moore, P.D. and Webb, J.A. (1978) *An Illustrated Guide to Pollen Analysis* (Hodder and Stoughton: London)

Mosley, M.P. (1981) The influence of organic debris on channel morphology and bed load transport in a New Zealand Forest Stream, *Earth Surface Processes and Landforms* 6: 571–580

Moss, R.P. (1972) Seven pillars of wisdom: a note on methodology, *Area* 4: 237–241

Moss, R.P. (1977) Deductive strategies in geographical generalization, *Progress in Physical Geography* 10: 23–39

Moss, R.P. (1979) On geography as science, *Geoforum* 10: 223–233

Munn, R.E. (1979) *Environmental Impact Assessment*, Scientific Committee on Problems of the Environment, 5 (Wiley: Chichester)

Murdoch, W.W. (1966) Community structure, population control and competition – a critique, *The American Naturalist* 100: 219–226

Nagel, E. (1961) *The Structure of Science* (Harcourt, Brace and World Inc: New York)

Nash, J.E. and Sutcliffe, J.V. (1970) River flow forecasting through conceptual models. Part I – a discussion of principles, *Journal of Hydrology* 10: 282–290

Newton-Smith, W.H. (1981) *The Rationality of Science* (Routledge and Kegan Paul: London)

O'Connell, P.E., Nash, J.E. and Farell, J.P. (1970) River flow forecasting through conceptual models. Part II – the Brosna catchment at Ferbane, *Journal of Hydrology* 10: 317–329

Odum, E.P. (1969) The strategy of ecosystem development, *Science* 164: 262–270

O'Hear, A. (1980) *Karl Popper* (Routledge and Kegan Paul: London)

Ostrem, G., Liestol, O. and Wold, B. (1976) Glaciological investigations at Nigardsbreen, Norway, *Norsk Geografisk Tidsskrift* 30: 187–209

Overton, W.S. (1977) A strategy for model construction, in *Ecosystem Modelling in Theory and Practice*, A.S. Hall and J.W. Day (Wiley: New York) pp. 49–74

Palmer, J. and Radley, J. (1961) Gritstone of the English Pennines, *Zeitschrift für Geomorphologie* NF5: 37–52

Pardee, J.T. (1910) The glacial lake Missoula, *Journal of Geology* 18: 376–386

Peters, R.H. (1976) Tautology in Evolution and Ecology, *American Naturalist*, 110: 1–12

Pinder, G.F. and Jones, J.F. (1969) Determination of the groundwater component of peak discharge from the chemistry of total runoff, *Water Resources Research* 5: 438–445

Pitty, A.F. (1982) *The Nature of Geomorphology* (Methuen: London)

Platt, J.R. (1964) Strong inference, *Science* 146: 347–353

Playfair, J. (1802) *Illustrations of the Huttonian Theory of the Earth*, reprinted 1964 (Dover: New York)

Popper, K.R. (1962) *The Open Society and Its Enemies*, 2 volumes (Routledge and Kegan Paul: London)

Popper, K.R. (1970) Normal science and its dangers, in *Criticism and the Growth of Knowledge*, I. Lakatos and A. Musgrave (Cambridge University Press: Cambridge) pp. 51–58

Popper, K.R. (1972a) *The Logic of Scientific Discovery*, sixth revised impression (Hutchinson: London)

Popper, K.R. (1972b) *Objective Knowledge* (Oxford University Press: Oxford)

Popper, K.R. (1974) *Conjectures and Refutations*, fifth edition (Routledge and Kegan Paul: London)

Popper, K.R. (1976) *Unended Quest: An Intellectual Autobiography* (Fontana: London)

Popper, K.R. (1980) Evolution, *New Scientist* 87: 611

Popper, K.R. (1983) Realism and the aim of science, from the *Postscript to the Logic of Scientific Discovery*, W.W. Bartley III (Hutchinson: London)

Prestwich, J. (1886) *Geology: Chemical, Physical and Stratigraphical*, 2 volumes (Oxford)

Rich, J.L. (1911) Recent stream trenching in the semi-arid portion of Southwestern New Mexico, ·a result of reduced vegetation cover, *American Journal of Science* 32: 237–245

Ridley, M. (1981) Who doubts evolution? *New Scientist* 90: 830–832

Rodda, J.C. (1970) A trend-surface analysis trial for the planation surfaces of North Cardiganshire, *Transactions of the Institute of British Geographers* 50: 107–114

Rosen, E. (1959) *Three Copernican Treatises* (Dover: New York)

Rosenquist, M.A. (1955) *Investigations in the Clay-Electrolyte-Water System*, Norwegian Geotechnical Institute Publication No. 9

Ruse, M. (1981) Darwin's theory: an exercise in science, *New Scientist* 90: 828–830

Ruse, M. (1982) A philosopher at the monkey trial, *New Scientist* 93: 317–319

Russell, B. (1961) *A History of Western Philosophy*, second edition (George Allen and Unwin: London)

Rymer, L. (1978) The use of uniformitarianism and analogy in palaeoecology, particularly pollen analysis, in *Biology and Quarternary Environments*, D. Walker and J.C. Guppy (Australian Academy of Sciences: Canberra) pp. 245–257

SRE (1977) A non-conformist extract from a northern eyrie, *Area* 9: 122–123

Sandford, K.S. (1929) The erratic rocks and the age of glaciation in the Oxford district, *London Geological Journal*, 85: 359–388

Scheidegger, A.E. (1961) *Theoretical Geomorphology*, first edition (Springer: Berlin)

Schilpp, P.A. (1974) *The Philosophy of Karl Popper* Library of Living Philosophers XIV, 2 volumes (Open Court: La Salle, Illinois)

Schoener, T.W. (1976) The species–area relation within archipelagos: models and evidence from island biotas, *Proceedings of the Sixth International Ornithological Congress, Canberra, August 1974*

Schumm, S.A. and Lichty, R.W. (1965) Time, space and causality in geomorphology, *American Journal of Science* 263: 110–119

Selby-Bigge, L.A. (1902) (editor) *Enquiries Concerning Human Understanding and Concerning the Principles of Morals* by David Hume (Clarendon Press: Oxford)

Shapere, D. (1964) The structure of scientific revolutions, *Philosophical Review* LXXIII: 383–394

Shea, J.H. (1983) Twelve fallacies of uniformitarianism, *Geology* 10: 455–460

Shreve, R.L. (1975) The probabilistic approach to drainage basin geomorphology, *Geology* 3: 527–529

Siegel, S. (1956) *Non-parametric Statistics for the Behavioural Sciences* (McGraw Hill: London)

Simberloff, D. (1976) Experimental zoogeography of islands: effects of island size, *Ecology* 57: 629–648

Simberloff, D. and Wilson, E.O. (1970) Experimental zoogeography of islands. A two-year record of colonisation, *Ecology* 51: 934–937

Simpson, G.G. (1962) Historical science, in *The Fabric of Geology*, C.C. Albritton (Addison-Wesley: Reading, Mass.) pp. 24–48

Sissons, J.B. and Cornish, R. (1983) Fluvial landforms associated with ice-dammed lake drainage in upper Glen Roy, Scotland, *Proceedings of the Geologists' Association* 94: 45–52

Skolimowski, H. (1974) Karl Popper and the objectivity of scientific knowledge, in *The Philosophy of Karl Popper*, P.A. Schlipp, Library of Living Philosophers XIV, 2 volumes (Open Court: La Salle, Illinois) pp. 483–508

Sparks, J. (1981) What is this thing called science? *New Scientist* 89: 156–158

Sparks, J.J. (1962) Scientific method, *Bulletin of the Institute of Physics* November: 289–294

Sprout, B. (1984) Witch doctors and rainfall: the occult effect, *Journal of Bogus Science*, 12: 345–678

Stoddart, D. (1977) The paradigm concept and the history of geography. Abstract of paper for the conference of the International Geographical Union on the History of Geographic Thought (IGU: Edinburgh)

Strahler, A.N. (1950) Davis' concept of slope development viewed in the light of recent quantitative investigations, *Annals of the Association of American Geographers* 40: 209–213

Strahler, A.N. (1952) Dynamic basis of geomorphology, *Geological Society of America Bulletin* 63: 923–938

Strong, D.R. (1982) Null hypotheses in ecology, in *Conceptual Issues in Ecology* E. Saarinen (D. Rendell: Dordrecht)

Suppe, F. (1977) The search for philosophical understanding of scientific theories, in *The Structure of Scientific Theories*, F. Suppe (University of Illinois Press: Illinois) pp. 3–241

Tarrent, J.R. (1970) Comments on the use of trend-surface analysis in the study of erosion surfaces, *Transactions of the Institute of British Geographers* 51: 221–222

Taylor, P.J. (1975) An interpretation of the quantification debate in British geography, *Institute of British Geographers, Transactions* NS1: 129–142

Thomas, R.W. and Huggett, R. (1980) *Modelling in Geography: A Mathematical Approach* (Harper and Row: London)

Thorley, A. (1981) Pollen analytical evidence relating to the vegetation history of the Chalk, *Journal of Biogeography* 8: 93–106

Thornes, J.B. and Brunsden, D. (1977) *Geomorphology and Time* (Methuen: London)

Tinkler, J. (1975) *A Short History of Geomorphology* (Croom Helm: London)

Vincent, P. (1981) From theory to practice – a cautionary tale of island biogeography, *Area* 13: 115–118

Vincent, P. and Clarke, V. (1980) Terracette morphology and soil properties: a note on a cannonical study, *Earth Surface Processes* 5: 291–296

Vitousek, P.M. and Reiners, W.A. (1975) Ecosystem succession and nutrient retention: a hypothesis, *Bioscience* 26: 376–381

Walker, D. (1970) Direction and rate in some British postglacial hydroseres, in *The Vegetational History of the British Isles*, D. Walker and L.R. West (Cambridge University Press: Cambridge) pp. 117–139

Warner, R. (1954) *Thucydides: The Peloponnesian Wars* (Penguin Books: London)

Webb, B.W. and Walling, D.E. (1974) Local variation in background water quality, *The Science of the Total Environment* 3: 141–153

Wegener, A. (1912) *Die Entstehung der Kontinente* (Petermanns: Mitteilungen)

Wegener, A. (1966) *The Origin of Continents and Oceans*, translated from the fourth revised German edition of 1929 by J. Birman (Methuen: London)

Wheeler, P.B. (1982) Revolutions, research programmes and human geography, *Area* 14: 1–4

Wheeler, S.W.M. (1967) Foreword, in *Water Weather and Prehistory*, R. Raikes (Baker: London) p. iii

Whipkey, R.Z. (1965) Subsurface stormflow from forested slopes, *Bulletin of the International Association of Scientific Hydrology* 10: 74–85

Wiggins, I.L. and Porter, D.M. (1971) *Flora of the Galapagos* (Stanford University Press: Stanford)

Williamson, M. (1981) *Island Populations* (Oxford University Press: Oxford)

Wilson, A.J. (1972) Theoretical geography: some speculations, *Institute of British Geographers, Transactions* 57: 31–44

Wolman, M.G. and Miller, J.P. (1960) Magnitude and frequency of forces in geomorphic processes, *Journal of Geology* 68: 54–74

Wooldridge, S.W. (1958) The trend of geomorphology, *Institute of British Geographers, Transactions* 25: 30–35

Wooldridge, S.W. and Linton, D.L. (1933) The loam terrains of south east England and their relation to its early history, *Antiquity* 7: 297–310

Wooldridge, S.W. and Morgan, R.S. (1959) *The Physical Basis of Geography*, second edition (Longman: London)

Worsley, P. (1981a) Radiocarbon dating: principles, application and sample collection, in *Geomorphological Techniques*, A. Goudie (George Allen and Unwin: London) pp. 277–282

Worsley, P. (1981b) Lichenometry, in *Geomorphological Techniques*, A. Goudie (George Allen and Unwin: London) pp. 302–305

Wyllie, P.J. (1976) *The Way the Earth Works* (Wiley: New York)

Yates, F. (1970) *Experimental Design (Selected Papers)* (Griffin: London)

Zelinsky, W. (1974) The demigod's dilemma, *Annals of the Association of American Geographers* 65: 123–143

INDEX OF NAMES

SUBJECT INDEX